This book is a statewide study of Tennessee's agricultural population between 1850 and 1880. Relying upon massive samples of census data as well as plantation accounts, Freedmen's Bureau Records, and the Tennessee Civil War Veterans Questionnaires, the author provides the first systematic comparison of the socioeconomic bases of plantation and nonplantation areas both before and immediately after the Civil War. Although the study applauds scholars' growing appreciation of southern diversity during the nineteenth century, it argues that recent scholarship both oversimplifies distinctions between Black Belt and Upcountry and exaggerates the socioeconomic heterogeneity of the South as a whole. It also challenges several largely unsubstantiated assumptions concerning the postbellum reorganization of southern agriculture, particularly those regarding the impoverishment of southern whites and the immobilization and economic repression of southern freedmen.

One South or Many?

One South or Many?
Plantation Belt and Upcountry in Civil War–Era Tennessee

ROBERT TRACY McKENZIE

University of Washington

CAMBRIDGE
UNIVERSITY PRESS

FRANKLIN PIERCE
COLLEGE LIBRARY
RINDGE, N.H. 03461

Published by the Press Syndicate of the University of Cambridge
The Pitt Building, Trumpington Street, Cambridge CB2 1RP
40 West 20th Street, New York, NY 10011-4211, USA
10 Stamford Road, Oakleigh, Melbourne 3166, Australia

© Cambridge University Press 1994

First published 1994

Printed in the United States of America

Library of Congress Cataloging-in-Publication Data
McKenzie, Robert Tracy.
One South or many? : plantation belt and upcountry in Civil War–
era Tennessee / Robert Tracy McKenzie.
p.   cm.
Based on the author's thesis (doctoral – Vanderbilt University).
Includes index.
ISBN 0-521-46270-3 (hc)
1. Tennessee – Economic conditions.   2. Tennessee – Social
conditions.   3. Tennessee – History – Civil War, 1861–1864.
4. Reconstruction – Tennessee.   5. Agriculture – Economic aspects –
Tennessee – History – 19th century.   6. Plantations – Tennessee –
History – 19th century.   7. Farms, Small – History – 19th
century.   I. Title.
HC107.T3M28   1994
330.9768'04 – dc20   93-50235
                                                                    CIP

A catalog record for this book is available from the British Library.

ISBN 0-521-46270-3 hardback

HC
107
.T3
.M28
1994

For Robyn

Contents

Acknowledgments

Having devoted nearly eight years to this study, it seems as if I've been looking forward to writing these acknowledgments for most of my adult life. I've benefited immeasurably during these years from the instruction, guidance, encouragement, and support of colleagues, friends, and family members, and it is with long-anticipated pleasure that I now formally express my appreciation.

This book has evolved from a doctoral dissertation undertaken at Vanderbilt University, a school with a rich tradition in the study of rural and agricultural history that dates to the path-breaking work of Professor Frank L. Owsley and his graduate students during the 1930s and 1940s. My dissertation committee, in particular Professors Donald Winters and David Carlton, shaped this project in its earliest stages and inspired me to persevere in its completion by the high standards of professionalism and scholarship that each consistently demonstrates. I am privileged to consider both now as friends as well as mentors. The College of Arts and Sciences at Vanderbilt also supported my work financially, awarding me a dissertation improvement grant that paid for research trips both to the National Archives and to the Baker Library of Business Administration at Harvard. During my time in Nashville, I also was assisted enormously by Marilyn Bell and the fine staff at the Tennessee State Library and Archives, where I spent approximately two years sitting in the second microfilm carrel from the front in the row nearest the windows.

Since coming to the University of Washington I have been surrounded by numerous colleagues graciously willing to help a struggling new assistant professor. For their criticism as well as their praise, I am grateful to the History Department's History Research Group, especially members Bruce

Hevly, James Felak, John Findlay, Richard Johnson, and Richard Kirkendall, all of whom cheerfully read additional drafts of chapters or articles at my request. The staff of the Interlibrary Borrowing Service of Suzallo Library, Anna McCausland in particular, have also been vitally helpful, patiently handling my numerous requests for microfilm and making it possible through their efforts for me to continue the study of southern agriculture from the Pacific Northwest. The graduate school of the university also supported my work with two grants, the first in 1989 to finance a return trip to the Southeast and an uninterrupted summer of research, the second in 1993 to aid in the publication of this book. Additional travel to Tennessee was made unnecessary by a grant from the Howard and Frances Keller Endowed Fund in History that enabled the purchase of more than two dozen microfilm rolls of land and tax records.

Numerous individuals from other institutions have also given of their time and expertise to assist me, exemplifying the highest ideals of scholarship and collegiality thereby. Robert Higgs read almost the entire manuscript and responded rapidly with copious, penetrating remarks. Jonathan Atkins, Fred Bode, Peter Coclanis, Doug Flamming, Lacy Ford, Gordon McKinney, Mel McKiven, and Donald Schaefer each read drafts of chapters and provided valuable, constructive criticism. Vernon Burton, Carville Earle, Joseph Reidy, and Crandall Shifflett delivered comments on conference papers that have also proven helpful. Although I have not always heeded the suggestions of these scholars, the rare instances when I have not done so likely reflect my own intransigence more than the quality of their counsel. Needless to say, I alone am responsible for those flaws in the argument that remain.

Finally, with great emotion I recognize the invaluable contribution of my family. My grandfather B. L. Hale instilled in me early in life a love both of scholarship and of teaching. My greatest regret is that he passed away before I could complete this book, for I know he would have enjoyed bragging about his grandson, the published historian. My parents, Edwin and Margaret Lee McKenzie, have loved me unconditionally, supported me unequivocally, and modeled before me the virtues of industry and integrity. My daughters, four-year-old Callie and one-year-old Margaret, have no idea what daddy does at "the university" but have assisted me nonetheless by reminding me daily of the things that are truly important. And Robyn, my wife and best friend, has always believed in me. It is to her that I dedicate this book. *Song of Solomon 5:16*

Introduction

The internal diversity of the nineteenth-century South is simultaneously one of the most widely invoked and least explored themes in southern history. Whereas earlier generations were content to focus chiefly on the Black Belt and then extrapolate their findings to the entire South, since the 1970s scholars have been increasingly uncomfortable with such an approach, recognizing that it might exaggerate regional uniformity and lead to a distorted understanding of small-farm sections.[1] The result, first manifested extensively during the 1980s, has been a marked interest in the social and economic development of areas outside of the Black Belt, most notably the Upcountry areas of Georgia and South Carolina and the remote recesses of southern Appalachia.[2] Despite such recent work, however, our understanding of the socioeconomic bases of southern internal diversity is still primarily impressionistic. Like blind men groping an elephant, scholars

1. See, for example, Eugene D. Genovese, "Yeomen Farmers in a Slaveholders' Democracy," *Agricultural History* 49 (1975):331–42; Edward Pessen, "How Different from Each Other Were the Antebellum North and South?", *American Historical Review* 85 (1980):1119–49.
2. Primary examples include Steven Hahn, *The Roots of Southern Populism: Yeomen Farmers and the Transformation of the Georgia Upcountry, 1850–1890* (New York: Oxford University Press, 1983); David F. Weiman, "Petty Commodity Production in the Cotton South: Upcountry Farmers in the Georgia Cotton Economy, 1840–1880" (Ph.D. diss., Stanford University, 1983); Lacy K. Ford, *The Origins of Southern Radicalism: The South Carolina Upcountry, 1800–1860* (New York: Oxford University Press, 1988); Ronald D. Eller, *Miners, Millhands, and Mountaineers: Industrialization of the Appalachian South* (Knoxville: University of Tennessee Press, 1982); John Inscoe, *Mountain Masters, Slavery, and the Sectional Crisis in Western North Carolina* (Knoxville: University of Tennessee Press, 1989); and Robert D. Mitchell, ed., *Appalachian Frontiers: Settlement, Society, and Development in the Preindustrial Era* (Lexington: The University Press of Kentucky, 1991).

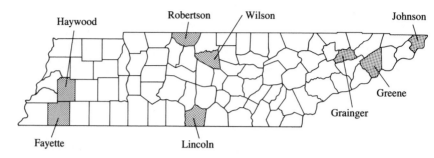

Sample Tennessee counties, 1860

have begun to describe different parts of the whole but as yet have no systematic basis for comparing them. Although scholars now frequently maintain that plantation and nonplantation areas differed in social and economic structure, explicitly comparative studies that explore such differences are rare.

In this book I have attempted to fashion such an explicitly comparative study by investigating and measuring the diversity of social and economic structure among rural Tennesseans during the era of the Civil War and Reconstruction. Perhaps no other state exhibited greater agricultural diversity on the eve of the Civil War than did Tennessee. As a federal report observed, "the length of the state . . . gives to the state its most prominent characteristic, to wit, great variety. This is seen," the report elaborated, "in its topography, geology, soil, climate, agriculture, and we may say in the character and habits of its population."[3] Although there are in actuality eight distinct geological regions within the state, for purposes of analysis the study will concentrate on the state's three "grand divisions," East, Middle, and West Tennessee. Not only is this approach practically simpler, but it also conforms to Tennesseans' own traditional views regarding the diversity of their state. Popular perceptions regarding its tripartite character are reflected in the three stars on the state flag and, until recently, in highway signs welcoming visitors to "the three states of Tennessee."[4]

To reduce the task to manageable proportions, I have focused particularly on the farm populations of eight counties. These eight counties are

3. Eugene W. Hilgard [Special Agent], *Report on Cotton Production in the United States* (Washington, DC: Government Printing Office, 1884), p. 383.
4. Robert E. Corlew, *Tennessee: A Short History*, 2d ed. (Knoxville: University of Tennessee Press, 1981), pp. 3–5; Paul H. Bergeron, *Paths of the Past: Tennessee, 1770–1970* (Knoxville: University of Tennessee Press, 1979), pp. 3–4.

shaded in gray on the map on page 2. Three of the counties selected for intensive analysis – Johnson, Greene, and Grainger – lie in the easternmost section of the state. The first white settlers entered upper East Tennessee – then part of the colony of North Carolina – at the end of the 1760s in defiance of the British Proclamation Line banning settlement west of the Appalachians. Typically following the Holston or French Broad rivers from Virginia or western North Carolina, settlers who entered the area found a region characterized by a succession of heavily wooded mountain ridges interspersed with narrow, generally fertile mountain valleys – "the Switzerland of America," in the words of local boosters.[5]

The southern Appalachians, known also as the Smoky Mountains, run along the border between Tennessee and North Carolina at an average elevation of 5,000 feet. Situated in the extreme northeastern tip of the state along this border, Johnson County, the first eastern county to be studied, is essentially a mountain county. Its cultivable portion consists of a long, straight valley running northeast to southwest that is bounded by Iron and Stone mountains. The valley, known as Johnson County Cove, is approximately thirty miles long and three to four miles across at its widest point, and it has an average elevation of about 2,000 feet. As the state's Commissioner of Agriculture observed of the cove in the 1870s, agriculturally "it *is* Johnson County."[6]

The mountain farmers of Johnson County were probably as close to total isolation from the surrounding region as it was possible to be in the nineteenth-century South. The county had a few tolerably good roads but no macadamized turnpikes. No railroad penetrated the county until the mid-1890s.[7] To enter or leave the cove it was necessary either to climb over the surrounding mountains or to pass through one of the narrow gaps cut through by mountain streams. To a considerable extent these obstacles effectively prevented involvement in a larger regional economy. The coun-

5. Corlew, *Tennessee: A Short History*, pp. 43–51; Hermann Bokum, *The Tennessee Handbook and Immigrants' Guide* (Philadelphia: J. B. Lippincott & Co., Publishers, 1868), p. 8.

6. Joseph Buckner Killebrew, *Introduction to the Resources of Tennessee* (Nashville, TN: Tavel, Eastman, & Howell, 1874; reprint ed., Spartanburg, SC: The Reprint Company, 1974), p. 543, emphasis added. See also *Goodspeed History of Tennessee* (Nashville: The Goodspeed Publishing Company, 1887), pt. 2, p. 922; Hilgard, *Report on Cotton Production*, p. 411; James M. Safford, *Geology of Tennessee* (Nashville, TN: S. C. Mercer, 1869), pp. 49–50.

7. William H. Nichols, "Some Foundations of Economic Development in the Upper East Tennessee Valley, 1850–1900," *Journal of Political Economy* 64 (1956):277–302.

ty's mountain farmers concentrated on livestock grazing and also grew corn, wheat, and oats, primarily for local consumption.

To the west of Johnson County lies the prosperous Valley of East Tennessee, so called because it is framed on the east by the Smokies and on the west by the Cumberland Tableland, a massive plateau approximately 2,000 feet above sea level. In the nineteenth century the East Tennessee Valley was one of the most prominent small-farm regions of the Upper South. The soil is generally quite fertile, and it is well suited to the production of most cereals and grasses. As a census official described the region in the 1880s, it was "the poor man's rich land."[8]

Grainger County lies wholly within the Valley, as does Greene, with the exception of a three- to six-mile strip of mountain land on its southeastern border. During the 1850s the East Tennessee Valley rivaled the Shenandoah as the breadbasket of the South; accordingly, farmers in both counties concentrated heavily on wheat as well as on corn and livestock. In each county farmers were limited by a lack of good roads but were reasonably well served by tributaries of the Tennessee River – the Clinch, Holston, and Nolichucky rivers. None of these was adequate for reliable steamboat navigation, but farmers used them to transport large quantities of produce by flatboat to the Tennessee River, and via the Tennessee to Knoxville, the leading agricultural market in the region. In addition, the completion in 1858 of the East Tennessee and Virginia Railroad directly connected farmers in Greene County not only with Knoxville but with Chattanooga, Atlanta, and the Deep South.[9]

None of the eastern counties depended significantly on slave labor for the production of agricultural goods; in East Tennessee as a whole slaves constituted but 8 percent of the population. This was not due primarily to any principled public opposition to the "peculiar institution." A small but vocal abolition movement did develop in the region – one of the earliest antislavery newspapers in the country was published in Greene County – but in East Tennessee as throughout the South the strength of antislavery sentiment peaked by the late 1820s and was virtually extinct by the following decade.[10] Rather, the relatively slight contribution of slave labor was

8. Hilgard, *Report on Cotton Production*, p. 410.

9. Killebrew, *Introduction to the Resources of Tennessee*, pp. 276–86, 487–500; *Goodspeed History of Tennessee*, p. 881; Nichols, "Some Foundations of Economic Development," p. 284.

10. The newspaper was Benjamin Lundy's *The Genius of Universal Emancipation*, published in Greeneville between 1822 and 1824. See Corlew, *Tennessee: A Short History*, pp. 215–17.

determined chiefly by the area's topography, soil, and climate, which precluded the successful cultivation of any of the South's most profitable staples: cotton, sugar, or tobacco. Forced to bid for slaves against planters and farmers who reaped large profits from these crops, most would-be slaveowners in East Tennessee depended on free labor instead. In the sampled eastern counties only one in ten farmers was a master as well.

Because of the obstacles to large-scale commercial agriculture in East Tennessee, the predominant flow of migration into the state began to bypass the region by the early nineteenth century. White migrants had begun to settle in Middle Tennessee as early as the 1780s, and within a few decades the region had surpassed the eastern section of the state in population, wealth, and political influence. Geographically, the region features the gently undulating Central Basin, which is surrounded on all sides by the Highland Rim averaging 1,000 feet above sea level. Agriculturally, the area was distinguished before the Civil War by the great variety of its crops and by the fine quality of its livestock. Famous for its horses, sheep, and mules, the area, as the state's commissioner of agriculture claimed with only mild hyperbole, "probably [had] as much fine stock as all the cotton states put together." Although the northernmost line of feasible cotton production ran through the region, few Middle Tennesseans planted the crop during the antebellum period. At the same time, however, a significant proportion did rely on the labor of black slaves, who constituted more than one-fifth of the section's population; in the three counties sampled from the area more than one-third of white farmers owned slaves. Taken together, these traits made Middle Tennessee a rarity in the antebellum South – an area committed simultaneously both to slave labor and to the extensive production of *foodstuffs* for the market.[11]

The first of the three counties sampled from the region, Wilson County, lies wholly within the Central Basin, "the Garden of Tennessee." Its clay and lime-based soil was admirably suited to the production of corn, wheat, and oats, and the grasses of its uplands and hills made excellent pasturage for livestock. A well-developed series of macadamized roads, as well as access via the Cumberland River to nearby Nashville and the railroad network that centered there, provided Wilson County farmers with convenient links to external markets.[12]

11. Killebrew, *Introduction to the Resources of Tennessee*, pp. 2–3, 619–24; Stephen V. Ash, *Middle Tennessee Society Transformed, 1860–1870: War and Peace in the Upper South* (Baton Rouge: Louisiana State University Press, 1988), pp. 2–12.

12. In addition, Wilson County was connected to Nashville by rail in 1871 when the

The second county, Lincoln, is situated approximately sixty miles to the southwest along the Alabama border. About two-thirds of the county lies within the Central Basin; much like Wilson County, this part of Lincoln County is marked by extremely fertile valleys broken by wooded hills and ridges. The remaining third, an eight-mile strip along its southern boundary, more properly belongs to the Highland Rim and contains soil of poorer quality. Although more likely to grow cotton than farmers farther north, Lincoln County farmers, like Middle Tennesseans generally, focused primarily on corn, wheat, and livestock. They were well connected to regional markets by the Elk River, which bisected the county before emptying into the Tennessee, and by a branch of the Nashville and Chattanooga Railroad, which provided access to East Tennessee and the Lower South.[13]

The final Middle Tennessee county, Robertson, lies seventy-five miles due north of Lincoln County within the northern Highland Rim along the Tennessee–Kentucky border. Although not quite so fertile as the soils of the Central Basin, Robertson's lands dependably produced large crops of corn, oats, wheat, and tobacco throughout the antebellum period. No major navigable river runs through the county, but the county was connected to Nashville during the 1850s by the Edgefield and Kentucky Railroad.[14]

The economy of Middle Tennessee had already reached a fairly advanced state of development by the time the first white settlers were beginning to penetrate the westernmost section of the state. Defined as the area stretching between the Mississippi River and the western course of the Tennessee, West Tennessee, or the Western District as it was then known, was occupied exclusively by the Chickasaw Indian Nation until 1819, at which time the state secured title by treaty. The bulk of the region consists of a relatively flat, broad plain that slopes gently toward the alluvial lands along the banks of the Mississippi. Although farmers in the northern half of the region concentrated on corn, tobacco, and livestock production, farmers in the southwestern corner of the state quickly settled on cotton as the predominant staple; by midcentury the region was producing four-fifths of the

Tennessee and Pacific Railroad was completed as far as Lebanon, the county seat. See Killebrew, *Introduction to the Resources of Tennessee*, pp. 1004–12; *Goodspeed History of Tennessee*, p. 345.

13. Killebrew, *Introduction to the Resources of Tennessee*, pp. 799–807; Corlew, *Tennessee: A Short History*, p. 204.

14. Killebrew, *Introduction to the Resources of Tennessee*, pp. 890–900.

state's cotton crop and had emerged as one of the leading cotton-producing sections in the Upper South.[15]

Fayette and Haywood, the two western counties selected for analysis, lie adjacent to each other in this latter section. The soil of both counties belongs to the Brown Loam Tablelands, a belt of soil that runs far into Mississippi and that sustained the production of some of the finest Mississippi Upland cotton. It is no accident, then, that their agricultural economies more closely resembled that of northern Mississippi than of the remainder of Tennessee. Approximately nine out of ten farm operators in the counties planted cotton; fully eight out of ten also owned slaves, who constituted two-thirds of the population. The devotion of Haywood and Fayette farmers to the cotton and slave economy was strengthened by their proximity to Memphis, which had emerged by the late antebellum period as the leading inland cotton center of the South. Although roads in both counties were uniformly poor, river and rail alternatives were plentiful. Haywood's farmers had access to the city via the Hatchie and Forked Deer rivers and, after the mid-1850s, by the Memphis and Ohio Railroad. No major river connected Fayette to Memphis, but two railroads traversed the county en route to the city (the Memphis and Ohio and the Memphis and Charleston), and it is literally true that every farm in the county lay within a few miles of one or the other.[16]

In sum, the eight counties that I have selected for analysis varied significantly on the eve of the Civil War, most notably with regard to the twin elements that have come to define the antebellum southern economy in the popular mind: slavery and cotton. Without anticipating unduly the conclusions of such a comparison, it might be worthwhile at this point to consider briefly the potential benefits of comparing the farm populations of such disparate areas. Specifically:

1. What major historical themes should be illuminated?
2. What types of insight with regard to these themes might reasonably be expected?
3. To what degree should such insights be broadly applicable to the South as a whole?

The preeminent theme of interest is so obvious that I repeat it at the risk of redundancy: From beginning to end, the analysis to follow is designed

15. Corlew, *Tennessee: A Short History*, pp. 228–9; Killebrew, *Introduction to the Resources of Tennessee*, pp. 1014–22.
16. Hilgard, *Report on Cotton Production*, p. 391; *Goodspeed History of Tennessee*, p. 346; Killebrew, *Introduction to the Resources of Tennessee*, pp. 1062–8.

explicitly to investigate the nature and extent of interregional diversity in the nineteenth-century South. If, however, the past is truly a "seamless web" – and I believe that it is – and interregional diversity was of crucial significance to the nineteenth-century South – and I believe that it was – then the question of interregional diversity should intersect with other issues of historical importance for the region. In particular, any advance in knowledge with regard to the South's heterogeneity should also affect our understanding of the region's extent of distinctiveness and discontinuity, questions that have divided scholars for decades.[17] A narrow focus that ignored such interrelationships would be a product of choice, not necessity.

I must make clear at the outset that I have consciously made such a choice by determining to address one but not both of these inextricably related questions. Focusing on the years both immediately before and after the Civil War, this study confronts squarely the issue of discontinuity between the Old and the New South; some of the work's most valuable insights pertain directly to the question. Both Souths were agriculturally and socioeconomically variegated, and a more informed recognition of this fact greatly facilitates the evaluation of change over time. I have resolved to remain silent, on the other hand, concerning the implications of my findings for the debate over southern distinctiveness. With due appreciation for the high quality of much that has been written on the subject, I remain convinced that the current debate is fueled most of all by theoretical and ideological differences and would be little affected by the empirical evidence I might offer.

With regard to the latter two questions – that is, the kinds of insights to be expected and their broader applicability – two attributes of the study are of crucial importance. The first concerns the nature of the evidence that it employs. With rare exceptions – the occasional diary or account book of the wealthy planter – nineteenth-century farmers left no intimate artifacts for the twentieth-century historian. Thus, although the analysis takes into account traditional forms of evidence whenever possible – plantation accounts, reports of Freedmen's Bureau agents, reminiscences of Tennessee veterans – the bulk of the evidence is *quantitative*: impersonal

---

17. For the most recent historiographical treatments of these themes, see John B. Boles and Evelyn Thomas Nolen, eds., *Interpreting Southern History: Historiographical Essays in Honor of Sanford W. Higginbotham* (Baton Rouge and London: Louisiana State University Press, 1987), in particular the essays by Drew Gilpin Faust, "The Peculiar South Revisited: White Society, Culture, and Politics in the Antebellum Period, 1800–1860," pp. 78–119; and Harold Woodman, "Economic Reconstruction and the Rise of the New South, 1865–1900," pp. 254–307.

numerical data on landownership and crop production patterns drawn from federal and local records. This single factor unavoidably constrains the range of issues to be examined, although not as severely as would be indicated by Arthur Schlesinger Jr.'s famous dictum, "almost all important questions . . . are *not* susceptible to quantitative answers."[18] One may reject Schlesinger's declaration as extreme (and more than a little defensive) while still recognizing that all historians must labor with evidence that is invariably imperfect and incomplete, and that wise scholars work within the limitations of their sources.

Keeping in mind the character of the evidence will enable the reader to understand better the focus of the analysis that follows. It concentrates on the distribution of land and slaves more than on perceptions of class consciousness, stresses patterns of wealth accumulation more than attitudes regarding social mobility, emphasizes the extent of market involvement more than individual feelings concerning economic independence or profit maximization. In short, it pays closer attention to structure and behavior than to consciousness, or *mentalité*. This is a relative generalization only, however. Although focusing on the structural and behavioral characteristics of the farm population, the study does not back away from the often glaring implications concerning the perception and motive of individual farmers. The local structure of wealthholding is, after all, the economic foundation of social consciousness, and behavior – the "only language that rarely lies"[19] – constitutes the most reliable interpreter of *mentalité* available to the historian of the nonelite.

In addition to the nature of the evidence, a second important attribute of the analysis concerns its research design. This is a *case study* or, more properly, a series of case studies. Authors of case studies typically feel great pressure to demonstrate the representativeness of their subjects and the broader significance of their findings. Neither editorial boards nor tenure review committees – nor the reading public, for that matter – are impressed by the intense examination of obviously idiosyncratic subjects. The understandable temptation, then, is for authors to overstate their case, to assert that some particular community or individual or organization was in actuality a perfect miniature of some larger historical universe. In this in-

18. Arthur Schlesinger, Jr., "The Humanist Looks at Empirical Social Research," *American Sociological Review* 27 (1962):768–71, emphasis in the original.
19. Edmund Burke, *Reflections on the Revolution in France* (Penguin ed., New York, 1982), quoted in John Patrick Diggins, "Comrades and Citizens: New Mythologies in American Historiography," *American Historical Review* 90 (1985):647.

stance, for example, it is tempting to maintain that the eight counties to be investigated were a microcosm of Tennessee and that Tennessee, which contained "nearly all the important physical and geological features of the states around it," was a microcosm of "the South."[20]

Neither claim will stand up, unfortunately. The first is rigorously unprovable; the sample counties do not represent Civil War–era Tennessee in any statistically verifiable sense. The second is logically indefensible, resting as it does on a denial of the internal diversity that the study is designed to explore. At bottom, any findings must resemble those of case studies generally – that is, they will be suggestive rather than conclusive. Even so, although a microcosm neither of the South nor of Tennessee alone, the sample counties – which stretched across 400 miles from the Appalachians to the Mississippi – did reflect vividly the heterogeneity that was a hallmark both of the state and of the South as a whole during the nineteenth century. As such, they constitute a fruitful proving ground on which to test several of the rather facile, frequently unsubstantiated assumptions now extant concerning southern diversity. The goal, then, is to challenge those generalizations, to stimulate other scholars to rethink and reformulate their conceptions of Black Belt and Upcountry, and in so doing to arrive at a fuller understanding of the many Souths of the Civil War era.

20. Hilgard, *Report on Cotton Production*, p. 383.

1. "The Most Honorable Besness in the Country": Farm Operations at the Close of the Antebellum Era

Alsey Bradford was only four years old when in 1826 his father, Hiram, decided to move his family from their home in Louisiana to the Forked Deer region of West Tennessee. A merchant, the elder Bradford was drawn northward by the economic potential of the lands newly acquired from the Chickasaw Indian Nation, lands only then beginning to open to white settlement. Hiram Bradford settled his wife and two sons in the recently organized county of Haywood and soon thereafter opened one of the first general stores in the county seat of Brownsville, a small village advantageously situated between the Big Hatchie and Forked Deer rivers. Growing up with the country, the younger Bradford followed in his father's footsteps, describing himself in 1851 (at age twenty-nine) as a "sort of merchant." Within a few short years, however, Alsey Bradford had adopted, as his father had before him, a different vocation. "I enjoy," he recorded in his journal in 1855, "the estimable privilege of tilling the soil as a planter."[1]

In turning to the land, Bradford turned to a way of life shared by most Tennesseans in the 1850s. Citizens of the Volunteer State, as in every southern state before the Civil War, were predominantly farmers. From the alluvial cotton lands along the banks of the Mississippi River to the mountain valleys of the Appalachians, an average of three out of four rural households earned their livelihood directly from the soil (see Table 1.1). Even this figure may underestimate the pervasive importance of the farm. Undoubtedly, some unknown proportion of household heads reporting

1. Bradford Family Papers, 1824–1900, Tennessee State Library and Archives [hereinafter cited TSLA]; *Goodspeed History of Tennessee* (Nashville: The Goodspeed Publishing Company, 1887), pp. 818–25.

Table 1.1 Occupations of free household heads, sample Tennessee counties, 1860 (percentages)[a]

	East	Middle	West
Farmers and farm laborers[b]	75.1	75.4	80.0
Skilled laborers	10.0	10.3	5.6
Unskilled laborers	3.6	2.4	1.3
White collar and commercial	2.0	2.3	3.9
Professional	1.9	2.8	3.9
Miscellaneous and none given	7.6	6.9	5.4

[a]Sum of percentages may not equal 100.0 due to rounding. [b]Includes overseers.
Source: Eight-county sample, see text.

nonagricultural occupations to the local census enumerator inadvertently concealed significant ties to the soil.[2] Perhaps this was unavoidable, for in the local economies of rural nineteenth-century America the line between agricultural and nonagricultural sectors was blurred to an extent that twentieth-century Americans can scarcely comprehend. Although census agents rarely recorded dual occupations for their respondents – indeed, they were instructed not to – the recollections of Civil War veterans from the state make clear that antebellum Tennesseans were frequently shoemakers and farmers, carpenters and farmers, lawyers or teachers or preachers and farmers.[3]

2. Many household heads reported nonagricultural occupations but were also listed on the census of agriculture as farm operators. Other households headed by non-farmers included one or more members who reported agricultural occupations. When the definition of farm households is expanded to include both categories, the proportion of the free population residing in farming households in the eight sample counties was 84.7 percent in East Tennessee, 82.7 percent in Middle Tennessee, and 80.2 percent in West Tennessee.

3. Official instructions to census enumerators are recounted by Fred A. Bode and Donald E. Ginter in *Farm Tenancy and the Census in Antebellum Georgia* (Athens: University of Georgia Press, 1986), pp. 47ff. Between 1915 and 1922 the Tennessee Department of Archives and History collected responses to detailed but crude sociological questionnaires from 1,648 surviving Civil War veterans then living within the state. The responses are preserved in Gustavus W. Dyer and John Trotwood, comps., *The Tennessee Civil War Veterans Questionnaires*, 5 vols. (Easley, SC: Southern Historical Press, 1985). For references to multiple occupations, see, among many, vol. 1, pp. 71, 108–9, 283–4; vol. 2, pp. 460–1, 541–2, 560–2, 586–7, 595–6, 698–700, 731–2; vol. 3, pp. 877–9, 924–5, 975–6, 1019–20. Fred A.

This widespread dependence on the land was accompanied by an equally prevalent veneration of the land and of the landowner, the independent farmer whom Tennesseans identified as the bulwark of a Jeffersonian republic. Adhering to the tenets of what one scholar has labeled "country republicanism," the state's common folk regarded agriculture as both "the master sinew of every great State" and "the most natural and innocent employment of man."[4] All across the state, farmowners – as well as those who aspired to own farms – would have echoed Alsey Bradford's opinion, though often less eloquently, regarding the "privilege of tilling the soil." Reflecting decades later, Middle Tennessean Joseph Cardwell observed that before the Civil War "farming was an umble but exalted occupahan." East Tennessean M. V. Jones agreed; in the mind of this Civil War veteran, "farmming was [the] most honorable besness in the country."[5]

Yet, if farmers throughout the state shared a common bond in their ties to the land and in the dignity they accorded their role in society, they differed significantly in the routine and material quality of existence that the land provided them. Alsey Bradford, for example, prospered mightily as a "tiller of the soil," so much so that by 1860 he owned four tracts of land totaling 1,850 acres valued at nearly $20,000. Equally impressive, he had acquired by that time a total of thirty-one slaves; with their labor and that of nineteen draft animals, Bradford's home plantation, "Hatchie Heights," yielded 2,000 bushels of corn and more than 100 bales of cotton in 1859. On the other hand, although they held similar convictions regarding their vocations, both M. V. Jones and Joseph Cardwell extracted a far more modest living from the land. On the eve of the Civil War, Jones labored for his father, who owned five slaves and a small farm in East Tennessee not far from the Georgia border. With his son's help, B. F. Jones planted corn, wheat, oats, and potatoes on his 100 improved acres and kept meat and butter on the table by raising an assortment of livestock, including three milk cows, seven beef cattle, twelve sheep, and twenty-eight hogs. Although twenty-six years old on the outbreak of war, Joseph

Bailey analyzes the responses in *Class and Tennessee's Confederate Generation* (Chapel Hill: University of North Carolina Press, 1987).

4. *The Tennessee Farmer* I (1834):46. On country republicanism see Lacy K. Ford, Jr., *Origins of Southern Radicalism: The South Carolina Upcountry, 1800–1860* (New York: Oxford University Press, 1988), pp. 49–51. On the central role of republicanism in the political culture of Tennessee, see Jonathan Moore Atkins, " 'A Combat for Liberty': Politics and Parties in Jackson's Tennessee, 1832–1851" (Ph.D. diss., University of Michigan, 1991), Chaps. I and IV.

5. Dyer and Moore, comps., *Tennessee Veterans Questionnaires*, vol. 3, p. 1258; vol. 2, p. 446.

Cardwell, like M. V. Jones, also worked for relatives as a farm laborer. "No[t] one foot of land did I posses," the Middle Tennessean recalled years later, "nor a cottage in the wilderness."[6]

Much of the great variation among the state's farmers is explained by the geographical diversity of the state itself. Farm operations of the scale undertaken by Alsey Bradford were common in southwest Tennessee, a region that closely resembled the alluvial lands of northern Mississippi. They were highly unusual elsewhere in Tennessee, however; the sheer magnitude of Bradford's activities distinguished him from at least nine-tenths of the state's farmers. A comparison of Bradford's Tennessee to the Tennessee of the Central Basin or the East Tennessee Valley forcefully confirms Gavin Wright's observation that "the South never had much homogeneity on a purely geographic basis, even in the days when agriculture predominated."[7]

Even so, there was more than just geographical differentiation behind the diversity of farm experiences across the state. The remaining 90 percent of farmers who failed to match Bradford's level of output and prosperity were far from homogeneous; furthermore, regional generalizations are wholly inadequate to convey their dissimilarity. Indeed, small farmers differed as widely among themselves as they did from the large cotton planters in major plantation counties such as Fayette or Haywood.

This chapter analyzes patterns of farm operations in the sample Tennessee counties and seeks to capture a measure of the agricultural diversity that characterized the state and the South as a whole during the late antebellum period. It examines patterns of slaveholding, the work routine of everyday life, the overall scale of farm operations, the structure of the farm community, the degree of self-sufficiency and commercial orientation among individual farmers, and the level of income that their farms provided. An analysis of this sort should do much to highlight and evaluate the heterogeneity that was a prominent feature of southern agriculture on the eve of the Civil War.

Patterns of Slaveownership

Traveling across the state from northeast to southwest, visitors to Tennessee in the 1850s could only have been impressed by its internal diversity.

---

6. Bradford Family Papers, TSLA; Dyer and Moore, comps., *Tennessee Veterans Questionnaires*, vol. 3, p. 1258; vol. 2, p. 446; Eighth Agricultural Census [1860]; Duke University Library, Microcopy M6272, reels 73, 74.
7. Gavin Wright, *Old South, New South: Revolutions in the Southern Economy Since the Civil War* (New York: Basic Books, Inc., 1986), p. 5.

Beginning their excursion in the mountainous "Switzerland of America," then passing through the prosperous mixed-farming "garden" of the state – the Central Basin centered around Nashville – they would have concluded their journey in Memphis, the inland commercial center of the cotton frontier. The trip presented a striking study in contrasts, and a discerning eye would have observed literally countless reasons for Tennesseans' own perceptions of a state distinctly divided into three "grand divisions."[8] Of the many obvious manifestations of diversity, however, none was likely more striking than the increasing importance of slavery to the agricultural economy as the trek progressed westward.

The eight counties selected for this study reflect the marked variation in the economic significance of slavery that characterized the state as a whole. In the eastern counties barely one-tenth (10.4 percent) of free farm household heads owned one or more slaves.[9] For the bulk of slaveless whites in such areas, intimate daily contact with either slaves or slaveholders was rare. East Tennessean John Cooley, for example, could remember only five or six men in his neighborhood who owned slaves. Another Civil War veteran from the region, Andy Guffee, could not recall a single slaveholder within "more than five miles of where I lived." In the middle counties, in contrast, over one-third (36.3 percent) of free families owned slaves, and in the western counties almost two-thirds (62.4 percent) of farmers were masters. Haywood County veteran William Graves estimated that 80 percent of whites in the vicinity of his home "were worked for by others." Graves's statistical sense was uncanny. Considering farmowners only, almost exactly four-fifths of those who owned land in Haywood and Fayette counties owned human property also. Clearly, in such regions the presence of the institution was ubiquitous and its influence pervasive.[10]

Actually, figures regarding the extent of slaveownership understate the disparity among the regions in the importance of the "peculiar institution." Not only did the proportion of white farmers owning slaves increase from east to west, but the average number of slaves per slaveowner increased markedly as well. The median number of slaves per master was only three in East Tennessee; in that region approximately three out of five owners (61.4 percent) held fewer than five slaves, and only one in fifty (1.8 percent)

8. Paul H. Bergeron, *Paths of the Past: Tennessee, 1770–1970* (Knoxville: University of Tennessee Press, 1979), pp. 3–4.

9. Unless otherwise noted, the quantitative conclusions in this chapter are based on an analysis of a random, tenure-stratified sample of 4,015 farm household heads drawn from the eight sample counties. See Appendix A.

10. Dyer and Moore, comps., *Tennessee Veterans Questionnaires*, vol. 2, pp. 553–4; vol. 3, pp. 972–3, 946–8.

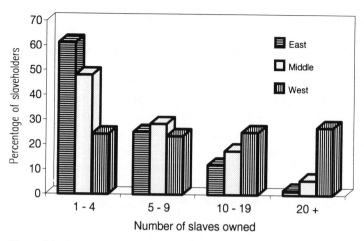

Figure 1.1 Patterns of slaveholding among free farm households, sample Tennessee counties, 1860

owned as many as twenty, thus meeting the standard definition of a planter (see Figure 1.1). In the central counties the median number of slaves per owner increased to five; larger slave holdings were more common, yet planters still constituted only 6 percent of all slaveowners. In the western district by comparison, the median number of slaves per master jumped to ten, and over one-quarter of all owners (27.4 percent) held a minimum of twenty bondsmen. Perhaps the best overall indicator of the agricultural importance of slavery is the number of slaves directly engaged in agriculture per white farm operator. Considering slaves of all ages and both sexes, there was less than one slave (0.9) per farm operator in the sample eastern counties, 3.4 slaves per operator in the middle counties, and 14.3 slaves per operator in the sample counties of western Tennessee.[11]

Work Routine

Theoretically, differences of such magnitude in the extent and average scale of slaveownership should have contributed to other significant differences

11. Actually, the foregoing discussion (as well as the patterns in Figure 1.1) applies to slaves held within the county of enumeration rather than to total slaves owned. Enumerators likely grouped hired slaves with the current slave renter rather than with the owner. Slaves owned by local farmers but laboring in other counties were excluded entirely. Neither factor should affect significantly the general conclusions on regional differences; indeed, given the likelihood that masters owning slaves in multiple counties were more common in West Tennessee than elsewhere, the disparity among the regions was probably even larger than indicated.

in the daily routine of the farmer, the structure of the local farm population, and the contours of farm operations and production. For the individual slaveholder, the most immediate practical effect of slaveownership involved the degree to which the control of slave labor alleviated for himself or his family the monotonous toil that was a defining characteristic of agricultural life. West Tennessean Richard McGee, for instance, remembered his youth on a small eighty-acre farm in Dyer County as one of unrelenting toil. McGee "went all of the usual farm paces"; he "plowed, hoed, cleared land of timber, made rails and oak boards, pickets & pulled fodder, cut wheat and oats with a cradle, [and] broke young stock to harness and saddle." The demands on farm women were typically no less severe. East Tennessee veteran John Hoback recalled that his mother "spun, wove, carded, knit, cooked, washed, and [performed] all kinds of house work." In addition, as her son observed in an afterthought, Elizabeth Hoback "bore and raised 12 children."[12]

By acquiring even one slave, farmers could begin to lessen the burden on their wives and daughters. Women in slaveholding households were far less likely to do their own household chores than were the wives of slaveless husbands; as the number of slaves in the household increased the manual tasks of the mistress lessened and her managerial responsibilities grew.[13] Women in households that owned a few slaves often shared their domestic duties with a single female slave; in such cases the mistress might continue to do all the sewing and weaving while delegating the more onerous jobs, the cooking and milking and washing, for example, to her slave helper. Mistresses of large slaveholding households, on the other hand, primarily supervised the labors of their servants. Sophronia Greaves, for instance, whose planter husband owned fifty slaves, "over looked household affairs" but limited her direct contribution to "fancy sewing." Similarly, Harriet Webster, whose husband owned seventy-five slaves in Middle Tennessee, developed the reputation of being "an industrious and energetic house-keeper" with the aid of "2 cooks, 2 laundry women, and 2 maids."[14]

12. Dyer and Moore, comps., *Tennessee Veterans Questionnaires*, vol. 4, pp. 1440–2; vol. 1, p. 71.

13. Considering the mothers of the Tennessee questionnaire respondents, approximately nine-tenths of those in slaveless households performed their domestic chores themselves; in slaveholding households with under twenty slaves the proportion was approximately one-third, and in households with twenty or more slaves the proportion dropped to less than 8 percent. See Bailey, *Class and Tennessee's Confederate Generation*, p. 150; and James Oakes, *Slavery and Freedom: An Interpretation of the Old South* (New York: Alfred A. Knopf, 1990), pp. 94–6.

14. Dyer and Moore, comps., *Tennessee Veterans Questionnaires*, vol. 2, pp. 677–9; vol. 3, p. 947; vol. 5, p. 2151.

Slaveownership also reduced the dependence of the master on the labor of his sons and thus, at least potentially, lightened their work load as well. A minority of slaveless Tennesseans assumed that slaveownership guaranteed idleness for those who owned them. Marcus Wiks, whose father was a landless farm laborer in Middle Tennessee, believed that "the slave holders idled and there slaves done the work." A slaveless veteran from an adjacent county maintained that "the welthy onde nigroes to do ther work and did nothing." William Beard, son of a tenant farmer in the western part of the state, echoed these bitter feelings but focused specifically on the indolence of slaveowners' sons. "Slave holders boys," Beard believed, "did not work any. [T]hey would even make a negro slave hand him a drink of water when sittin in [the] house."[15]

Such views were obviously exaggerated and, significantly, not very widely shared, even among the poorest of antebellum Tennesseans, who typically believed that white men and boys, even those in slaveholding families, shouldered their fair share of the physical burdens of farm life.[16] If we can trust the recollections of Civil War veterans from slaveholding families, only the very largest slaveholders (i.e., those who owned well over twenty slaves) were substantially freed from extensive manual labor. These rare masters primarily supervised the labors of their bondsmen or, if they had the means, employed a white overseer for this task and devoted their own time to other pursuits. George Webster, for instance, "had good overseers" to coordinate the work of his seventy-five slaves, allowing him to spend "most of his time fishing and hunting." Middle Tennessee veteran James T. McColgan averred that his father, who owned fifty slaves and a plantation of 1,000 acres, did "very little if any" hard physical labor. The male white children in such households also frequently reaped the benefits of abundant black labor. The younger McColgan, for example, spent most of his time in school; he "sometimes may have worked in the garden," but he certainly "never ploughed or made a field hand." Similarly, an East Tennessee veteran whose father owned forty slaves admitted that "I did some farm work but not much." Creed Haskins, son of a large West Tennessee slaveholder, recalled simply that he "never worked when a boy."[17]

15. Ibid., vol. 1, pp. 136, 300; vol. 4, p. 1602.

16. Ibid., vol. 2, p. 678. Considering respondents from all socioeconomic ranks, 80.6 percent of Tennessee veterans maintained that both slaveholding and slaveless whites worked hard physically. Among the "poor," those from families owning no slaves and fewer than eighty acres of land, the proportion fell only to 74.5 percent. See Bailey, *Class and Tennessee's Confederate Generation*, p. 158.

17. Dyer and Moore, comps., *Tennessee Veterans Questionnaires*, vol. 5, p. 2151; vol. 4, p. 1425; vol. 1, p. 317; vol. 3, p. 1038; see also J. William Harris, "The Organ-

For most slaveholding families, however, slave labor served as a supplement to, rather than a substitute for, the toil of family members. The most convincing evidence in this regard is indirect. Potentially, at least, one of the advantages of slaveownership was that it freed fathers who owned slaves to send their sons to school for longer periods than otherwise would have been possible. An examination of school attendance patterns suggests, however, that with the exception of masters who owned twenty slaves or more, slaveholding fathers did not send their sons to school for significantly longer periods than did slaveless yeomen.[18] Excluding the sons of the wealthiest planters, white male children in slaveholding households almost always worked alongside their slaves in the fields. Calvin Crook, whose father owned five slaves and a small farm in central Tennessee, "worked every day both with plow and hoe side by side with slaves." A Middle Tennessean whose family owned fifteen slaves remembered plowing and hoeing in the fields with the slaves and concluded that "we did all kinds of farm work together." A veteran from East Tennessee noted more generally that when fathers "had slaves they worked them, but their sons, if they had any[,] also worked."

The son of a small West Tennessee slaveholder suggested implicitly that this generalization held most for the smaller slaveowners. In Robert Dew's community, "few families owned from 25 to 100 negroes" but "the great majority . . . owned a few"; the result was that "all worked together[,] whites and blacks alike."[19] To the degree that this rule was violated within communities, then, it was likely to be among the wealthiest slaveholding families, those with twenty or more slaves. Keeping in mind the varying patterns of slaveownership among the state's regions, it seems probable to conclude that East Tennessee masters rarely if ever escaped the physical demands of the farm and that Middle Tennessee masters did so with only slightly greater frequency. In West Tennessee, on the other hand, masters for whom physical toil was unnecessary, although never typical, would have been nonetheless far more common.

Scale of Operations

If the implications of slaveownership for the physical routine of the master are in some sense problematic, the impact of slaveownership on the scale

---

ization of Work on a Yeoman Slaveholder's Farm," *Agricultural History* 64 (1990): 39–52.

18. Bailey, *Class and Tennessee's Confederate Generation*, p. 152.

19. Dyer and Moore, comps., *Tennessee Veterans Questionnaires*, vol. 2, p. 595; vol. 1, p. 246; vol. 2, pp. 587, 678.

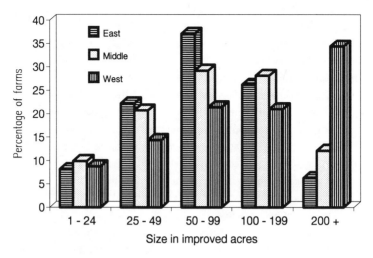

Figure 1.2 Farm size profile, sample Tennessee counties, 1860

of farm operations is obvious and direct. A crucial attribute of slavery was that it provided an elastic supply of labor to the individual farm operator, allowing an indefinite expansion of farm size. Consequently, the slaveholder could expand the scale of his operations far beyond that possible for the slaveless farmer, whose production was limited in theory by the amount of labor his family could provide.[20] An examination of mean farm size dramatically confirms this line of reasoning. East Tennessee farms averaged roughly 80 improved acres and Middle Tennessee farms 108, but West Tennessee farms were significantly larger, containing on average 206 improved acres. Apparently, then, the sheer magnitude of farm operations in the western counties sets them apart from the sample counties in the middle and eastern districts.

Despite huge disparities in the prevalence of slave labor, however, a closer examination reveals that the typical scale of farm operations varied far less than mean figures ostensibly suggest. To be precise, median farm size in the western counties was only 43 percent greater than in East Tennessee and 33 percent greater than in the central district. The median size was 70 improved acres in East Tennessee, 75 in Middle Tennessee, and only 100 in West Tennessee, less than one-half the mean level. As one

20. Gavin Wright, *The Political Economy of the Cotton South: Households, Markets, and Wealth in the Nineteenth Century* (New York: W. W. Norton & Company, 1978), pp. 44–55.

moved from east to west, in other words, the mean size of farms clearly increased at a much faster rate than did the median. This was the case, as common sense indicates and Figure 1.2 confirms, because as one moved westward the size distribution of farms became increasingly skewed in favor of large farms.

The advantages of slaveownership were not distributed evenly throughout the farming populations of each region. By definition, slave labor only relaxed the labor constraint of farmers who either owned or rented slaves; typically, farmers who operated units smaller than the median average for their regions were slaveless or, as was the case in West Tennessee, commanded the labor of so few slaves that they were unable to expand their operations greatly. Large slaveholders, on the other hand, invariably worked farms substantially above the median size. In differing degrees, the existence of such atypically large operators artificially inflated mean estimates of farm size in every region of the state. To cite some extreme examples, Fayette County's John W. Jones used 182 slaves to farm 2,000 improved acres – twenty times the West Tennessee median level – and to harvest more than 12,000 bushels of corn and 800 bales of cotton. Middle Tennessee tobacco planter George Augustus Washington, owner of 274 slaves, farmed 5,000 improved acres (sixty-seven times the regional median), which yielded in 1859 50,000 bushels of wheat, 22,000 bushels of corn, and 250,000 pounds of tobacco. Even in East Tennessee there were farm owners who, by the very size of their operations, towered above the local farm community. Grainger County's David Whiteside, for instance, owned twenty-two slaves and a plantation with 2,000 improved acres, roughly twenty-nine times the median size of farm units in the eastern counties.[21]

As the improved acreage distributions in Figure 1.2 make clear, farmers who operated on this scale increased substantially in number as one moved westward across the state. Units consisting of 200 or more improved acres constituted only 6.3 percent of all farms in East Tennessee. They were twice as prevalent in the Central Basin, where 12.1 percent of farms reached this size, and more than five times as common in West Tennessee, where one-third (33.4 percent) of all farm units were at least this large. At the other extreme, in contrast, the proportion of farmers working extremely small units (under twenty-five improved acres) was practically identical from region to region, varying from approximately 8 to 10 percent. As a consequence, no

21. Eighth Agricultural Census, Microcopy M6272, reels 73, 76, 74.

Table 1.2 Mean farm size in improved acres of farms above and below
the regional median, sample Tennessee counties, 1860

	East	Middle	West
Below regional median	40.8	42.2	54.4
Above regional median	124.3	182.0	372.4

Source: Eight-county sample, see text.

simple interregional comparisons regarding farm size are possible. Among
the bottom half of farm operators (those farming units containing fewer improved acres than the regional median), the average scale of farm operations
varied only modestly among the sections (see Table 1.2). Among the top half
of farm operators, they varied dramatically, primarily because of the abundant slave labor that western farmers commanded.

Despite the clear importance of slave labor, its uneven distribution
throughout the ranks of white farm operators cannot fully explain the relatively small interregional differences in median farm size across the state.
Regression analysis of the distribution of improved acreage across the entire
sample does show, as expected, that the number of slaves that an operator
owned was a powerful predictor of farm size. When a regression equation
is constructed that controls for the independent influence of region, however, as well as of tenure status, family size, and slave ownership, it is
apparent that farm units in East and Middle Tennessee were, other things
being equal, substantially larger than in the major slaveholding counties to
the southwest.[22] Thus, although slaveownership was apparently a crucial
factor in determining the feasible limits of farm production – at least it
was highly correlated with farm size – farmers in East and Middle Tennessee appear to have found ways partially to overcome their relative dearth
of slave labor. Including farms of all sizes, West Tennessee farms averaged
10.3 improved acres per household member (including slaves). In Middle
Tennessee per-capita improved acreage averaged 12.3 acres per farm, and

22. The equation is:

$$IA = 10.11 + 11.90S + 3.50F - 20.86T + 41.54E + 35.26M,$$
$$\quad\quad\;\;(71.1)\quad\;(4.1)\quad\;(-3.0)\quad\;(5.8)\quad\;(5.8)$$

where IA stands for improved acreage, S represents the number of slaves owned
by the farm operator, F equals the number of free persons in the operator's household, and T, E, and M are dummy variables for tenure, residence in East Tennessee, and residence in Middle Tennessee respectively. $R^2 = 0.714$; t-values are
in parentheses.

in East Tennessee the figure rose to 14.4 acres per household member, or a level approximately 40 percent greater than the West Tennessee average.

Two factors in particular may help to explain the apparent success of Middle and East Tennessee farmers in overcoming, however modestly, the labor constraint imposed by the relative shortage of slave labor in those regions. The first stems from the supposedly greater emphasis among eastern and central farmers on grazing and livestock husbandry. In the value of its livestock, Tennessee was always a leader among southern states during the antebellum era, and central and eastern Tennessee were always the leading regions within the state. Farmers in these regions may have evaded the labor constraint by giving proportionally greater attention to grazing, which required a relatively small input of labor, and less to the cultivation of staple crops, which by comparison was labor intensive. "Improved" acreage, as defined by the census, included any "cleared land used for grazing, grass, or tillage." Consequently, per-capita levels of improved acreage could have been greater in East and Middle Tennessee because farmers in those regions allotted relatively more acreage to grasses (red clover or timothy, for example) and relatively less to corn, wheat, oats, or tobacco.[23]

It is difficult to test this proposition because the 1860 agricultural census listed only the total number of improved acres, providing no information regarding the amounts of improved land planted in specific crops. Provided that yields of hay per acre planted in grass were roughly comparable from region to region, the ratio of total hay production (in tons) to total improved acres may serve as a crude indicator of the relative importance of grazing land as a component of total improved acreage. The number of tons of hay per improved acre equaled 0.034 in the eastern sample counties, 0.022 in the central counties, and 0.021 in the southwestern counties. Although as expected these figures show that eastern farmers did indeed place greater emphasis on hay production, the implied quantities of land allotted to grasses are quite small. Based on an estimated statewide yield of 1 1/4 tons per acre, it appears that eastern farmers sowed approximately 4 percent of their improved acreage in grasses, as compared to about 3 percent in the remaining two regions.[24] Clearly, the difference is too small to explain the substantial superiority of eastern farms in per capita improved acreage.

23. Census definitions are summarized by Bode and Ginter, *Farm Tenancy and the Census*, p. 58. For a discussion of the various grasses grown on Tennessee farms, see J. B. Killebrew, *Introduction to the Resources of Tennessee* (Nashville: Tavel, Eastman, & Howell, 1874; reprint ed., Spartanburg, SC: The Reprint Company, 1974), pp. 112–21.

24. Killebrew, *Introduction to the Resources of Tennessee*, pp. 104–5.

Structure of the White Farm Population

A second factor that might account for the greater per capita improved acreage among farmers in East Tennessee, at least, was the region's comparatively abundant supply of landless white labor. Many landless whites involved in agriculture were tenant farmers, that is, independent operators who rented their farms from others. Because the federal agricultural census before 1880 did not designate tenure status, historians have traditionally identified these individuals by matching farm operators enumerated on schedule IV of the federal manuscript census – the census of agriculture – to the comprehensive list of the free population recorded on schedule I. Scholars have reasoned, quite logically, that farm operators listed on schedule IV who reported no real wealth on schedule I were, with occasional exceptions, agricultural tenants.[25]

In all three regions of the state a substantial fraction of enumerated operators fit this description. The proportion of farm operators appearing on schedule IV who reported no wealth on schedule I – a good lower-bound estimation of the rate of agricultural tenancy – varied from approximately 17 percent in the eastern sample counties to 21 percent in the central counties to just under 14 percent in the southwestern plantation counties (see Table 1.3).[26] As these figures clearly demonstrate, tenants

25. This method was first employed by Frank L. Owsley and his graduate students at Vanderbilt University and later refined by Allan G. Bogue. See Owsley, *Plain Folk of the Old South* (Baton Rouge: Louisiana State University Press, 1949); and Bogue, *From Prairie to Corn Belt: Farming on the Illinois and Iowa Prairies in the Nineteenth Century* (Chicago: University of Chicago Press, 1963), pp. 56–66. For discussions of the method, see Robert P. Swierenga, "Quantitative Methods in Rural Landholding," *Journal of Interdisciplinary History* 4 (1983):795–9; Bode and Ginter, *Farm Tenancy and the Census*, pp. 33–44; and John T. Houdek and Charles F. Heller Jr., "Searching for Nineteenth-Century Farm Tenants: An Evaluation of Methods," *Historical Methods* 19 (1986):55–61.

26. Following Bode and Ginter's precise terminology, the lower-bound figures in Table 1.3 would be level II estimates of the tenancy rate. To ensure perfect comparability with their estimates for antebellum Georgia, I have purged the sample of farmers categorized by Bode and Ginter as types "b" and "c" (i.e., farmers reporting no crop production and little or no livestock). Unless explicitly defined otherwise, subsequent references to tenants are to "type a" farm operators on schedule IV (those with some arable production) who reported no real estate on schedule I. See *Farm Tenancy and the Census*, pp. 15, 112–13.

In addition, a comparison of real estate values reported on schedule I with farm values given on schedule IV suggests that between 6.1 and 8.0 percent of owner-operators in each region owned only a portion of their farms and rented the remainder (i.e., the value of their farm exceeded the value of their real wealth). Such operators might more properly be classified as tenants in part. If these operators

Table 1.3 Distribution of farm population, with lower- and upper-bound tenancy rates, household heads only, sample Tennessee counties, 1860

	East	Middle	West
Appearing on _both_ schedules I and IV			
(1) Owner-operators	2,521 (55.9)	3,956 (66.1)	1,409 (67.5)
(2) Definite tenants	507 (11.2)	1,056 (17.7)	224 (10.7)
Appearing on schedule I _only_			
(3) Landowners	198 (4.4)	283 (4.7)	153 (7.3)
(4) Landless	1,285 (28.4)	687 (11.5)	301 (14.4)
Estimated tenancy rate (percentage)			
Lower bound[a]	16.7	21.1	13.7
Upper bound[b]	39.7	29.1	25.2

[a]Equivalent to [(2)/(1)+(2)].   [b]Equivalent to {[(2)+(4)]/[(1)+(2)+(3)+(4)]}.
Source: Eight-county sample, see text.
Note: Figures in parentheses represent cell proportion of total households.

comprised a significant fraction of the agricultural populations of all three of the state's regions. Their material contribution to agricultural production was much smaller than their numbers would indicate, however, primarily because they typically farmed units that were substantially smaller than those of owner-operators (see Table 1.4). The proportion of total improved acreage farmed by tenants ranged from just over 5 percent in West Tennessee to almost 12 percent in East Tennessee. As a result, tenants generally

are classified as tenants rather than owners, the tenancy rates increase to 21.8 percent in East Tennessee, 27.3 percent in Middle Tennessee, and 20.6 percent in West Tennessee. It should be noted, however, that at least some of the discrepancies in "value of real estate owned" and "value of farm" were likely the result of enumerator error; also, some small proportion may have represented farm units jointly owned by two or more individuals. See Bode and Ginter, _Farm Tenancy and the Census_, p. 51; and Jeremy Atack and Fred Bateman, _To Their Own Soil: Agriculture in the Antebellum North_ (Ames: Iowa State University Press, 1987), pp. 110–1.

Table 1.4 Profile of farm size and farm operations by tenure, sample Tennessee counties, 1860

	Owner-operators		Tenants	
	Median value	Share of total (%)	Median value	Share of total (%)
East Tennessee				
Improved acres	70	88.3	50	11.7
Corn (bushels)	300	89.8	200	10.2
Cotton (bales)	NA	NA	NA	NA
Value of livestock	$400	91.1	$200	8.9
Middle Tennessee				
Improved acres	80	92.3	30	7.7
Corn (bushels)	600	89.9	250	10.1
Cotton (bales)	NA	NA	NA	NA
Value of livestock	$750	92.9	$250	7.1
West Tennessee				
Improved acres	120	95.0	40	5.0
Corn (bushels)	625	94.0	250	6.0
Cotton (bales)	21	94.7	7	5.3
Value of livestock	$870	94.8	$350	5.2

Source: Eight-county sample, see text.

accounted for between 5 and 10 percent of the agricultural output of their region.

Not all household heads engaged in agriculture appeared on the agricultural census, however, a fact that historians have long recognized but little explored (refer to Table 1.3). In every region a fraction (fewer than one in ten) of "farmers" who reported real wealth on schedule I were missing from the agricultural schedule (schedule IV). Scholars have variously hypothesized that such individuals were either farmowners who had just arrived in the community (and thus had no agricultural production to report for the previous year), retired farmers, or farmowners who rented out their entire farms to one or more tenants.[27]

27. On "farmers without farms" generally, see Merle Curti, *The Making of an American Community: A Case Study of Democracy in a Frontier County* (Stanford: Stanford

More crucial, a much larger proportion of farm household heads reporting no real wealth on schedule I also failed to appear on schedule IV. The function of these landless individuals, who were labeled as "farmers" or "farm laborers" but cannot be traced to a specific farm unit, is highly problematic. On the one hand, they may have been tenants who, for a variety of possible reasons, were omitted from schedule IV and whose acreage and production were grouped with that of the owner of the rented plot. For example, Frederick Bode and Donald Ginter, who have studied the intricacies of the antebellum census in great detail, speculate that some proportion of these landless farmers were sharecroppers – that is, laborers who worked a specific plot of land but were essentially employees of the landlord, receiving as wages a share of the crop that they helped to produce. Because sharecroppers, both in function and in legal status, resembled hired hands more than independent operators, Bode and Ginter hypothesize that at least some census marshals excluded them from the agricultural census and listed them on schedule I only.[28] Alternatively, these landless "farmers" and "farm laborers" may have been primarily unskilled agricultural laborers, individuals who worked by the day, month, or year as hired hands on farms operated by others.

Unfortunately, the actual occupational designations employed are not of much help in this regard because census officials appear to have used them in a highly inconsistent fashion. Landless individuals enumerated on the agricultural schedule often appear on schedule I as "farm laborers," for example, whereas landless household heads excluded from schedule IV are designated as "farmers."[29] Because enumerators' definitions of whom to include on the agricultural rolls likely varied greatly from county to county, the lower-bound tenancy estimates in Table 1.3 are not strictly comparable and may even present a seriously distorted impression regarding interregional variations. In contrast, the upper-bound estimates presented in the same table take into account not only the operators enumerated on schedule IV but also all landowning and landless farm household heads listed on schedule I only, some proportion of whom were unenumerated operators.

University Press, 1959), pp. 59–60; Bogue, *From Prairie to Corn Belt*, p. 64; Seddie Cogswell Jr., *Tenure, Nativity, and Age as Factors in Iowa Agriculture, 1850–1880* (Ames: Iowa State University Press, 1975).

28. Bode and Ginter, *Farm Tenancy and the Census*, pp. 90–113.
29. In Johnson County, for example, almost all nonowners were labeled farm laborers. In nearby Greene County, census agents in 1850 never employed the term farm laborer, and described almost all nonowners as farmers, even including the teenage sons of nonowning "farmer" fathers.

Although this second trio of figures undoubtedly exaggerates the tenancy rate in each region, they constitute precise estimations of total landlessness among the population of farm household heads, and unlike the lower-bound estimates, they may properly be used for interregional comparison.[30]

When "farmers" and "farm laborers" without farms are included in the computation of tenancy rates, a much greater disparity between East Tennessee and the other two sections emerges. As Table 1.3 shows, this is because the number of landless household heads omitted from the agricultural schedules was proportionally far higher in the eastern sample counties than elsewhere. Indeed, in East Tennessee, individuals in this rather ambiguous category constituted more than one-fourth of the entire free farm population, or more than double their proportion in the central and western districts.

At bottom, it would seem that a conclusive answer regarding the identity of this group is unattainable. In all likelihood, some proportion of the "farmers" and "farm laborers" without farms were, as Bode and Ginter theorize, sharecroppers, yet there is also indirect evidence that many were simply hired hands. For instance, in two of the counties with large numbers of farm household heads without farms (Grainger and Robertson), census enumerators simultaneously employed two different conventions in recording the acreage and production of landless operators on schedule IV, a practice which suggests that they were distinguishing between tenants and sharecroppers yet including both in the enumeration of farm operators.[31]

In addition, although there is not a single reference to antebellum sharecropping among the more than 1,600 Tennessee Civil War veteran questionnaires, there is a handful of isolated references to hired labor, most of them from East Tennessee. Very few veterans admitted that they or their fathers had been hired hands. Growing up in the East Tennessee Valley, William Patterson hired out to other farmers from the age of nine until he enlisted in the Confederate Army at the age of nineteen. In extreme upper East Tennessee, George Payne also hired out as a boy to neighboring farmers, as did his father, whom the son described as a "farm laverer by the day." More frequently, veterans recalled having adult hands on their families' farms. There were "generaly two men on [his father's] farm," one

30. These are equivalent to Bode and Ginter's level IV estimates. See *Farm Tenancy and the Census*, p. 113.

31. For some landless operators the enumerators recorded data for all relevant variables listed on the agricultural schedule, for others the enumerators omitted data on farm acreage and value while recording all other requested information, Bode and Ginter's "standard convention."

veteran observed, as well as "one girl or woman to help Mother." Similarly, an East Tennessean from Roane County related that his father "generally had two or three farm laborers who boarded at his house," and a veteran from a neighboring county recalled simply that his father "always had hired hands." Observing that his father "always had such working for him," the son of an East Tennessee slaveholder explained that, in his community at least, "the non–land owners hired as farm laborers."[32]

Historians have not seriously investigated the extent and importance of free white labor in the antebellum South. Rather, they have accepted uncritically what one prominent scholar has labeled "the overriding economic fact of life in early America[:] . . . the extreme difficulty farmers encountered in hiring non-family labor with which to expand their operations." Because farmers (1) preferred owning their own farms to working for others and (2) were typically able to acquire their own farms, the supply of landless free labor was understandably scarce.[33] Although there is eminent good sense in this axiom, and it explains much about agricultural development in both the North and the South, it has prompted historians to ignore subtle variations in the local availability of free agricultural labor.[34]

It is only technically an exaggeration to equate all nonfamily labor in the western sample counties with slave labor. If the British economist J. E. Cairnes exaggerated in defining as a "moral impossibility" the coexistence of slaves and free white laborers, he was essentially correct in observing that the former "render[ed] the capitalist independent" of the latter.[35] In the major plantation counties of West Tennessee the ratio of slave labor to

32. Dyer and Moore, comps., *Tennessee Veterans Questionnaires*, vol. 4, pp. 1705–6, 1710–1, 1519–22, 1435–7; vol. 2, pp. 541–2; vol. 4, p. 1670. See also Blanche Henry Clark, *The Tennessee Yeomen, 1840–1860* (Nashville: Vanderbilt University Press, 1942), pp. 18–19; Lewis Cecil Gray, *History of Agriculture in the Southern United States to 1860* (Washington, DC: Carnegie Institution of Washington, 1933), pp. 500–1. The forthcoming study of antebellum Tennessee agriculture by Donald Winters also testifies to the importance of white farm laborers in the state. I am grateful to Winters for sharing his findings while his work was still in press.

33. Wright, *The Political Economy of the Cotton South*, pp. 44–5.

34. In an otherwise flawed essay, Alfred Holt Stone observed over eight decades ago that scholarly emphases on the dearth of free white labor in the antebellum South resulted from the habit of "treating the geographic unit which we vaguely designate 'the South' as if it were characterized by absolutely uniform conditions throughout its length and breadth." See "Free Contract Labor in the Ante-Bellum South," in *The South in the Building of the Nation* (Richmond: Southern Historical Publication Society, 1909), vol. V, p. 140.

35. John E. Cairnes, *The Slave Power: Its Character, Career, and Probable Designs* (London: Macmillan & Company, 1863), 2d rev. ed., pp. 147, 143.

unattached white labor (landless whites not definitely tied to a specific farm)
was approximately 14:1. In Middle Tennessee the primacy of slave labor
was also strong, if less overwhelming; the ratio there was just above 4:1.
In East Tennessee, on the other hand, it was approximately 1:5. It is not
coincidental, then, that almost all the veterans who commented on the
importance of hired laborers grew up in East Tennessee. Nor is it an
accident that the rare references by veterans from other regions uniformly
lamented the limited opportunities in their communities for hired hands.
As a Middle Tennessee veteran observed, "slaveholders did not have to
higher mutch extry laber." A veteran from another slaveholding family
concurred: "As most of thoes in the community were slave holders There
was not much open for labor for any out sider. Only as bosses or overseers."[36] Although slavery provided an elastic supply of black labor to the
individual slaveowner, by limiting opportunities for landless whites the
prevalence of slaves in a given locality reduced the supply of white labor,
an alternative that flourished outside the Black Belt more than is recognized.[37]

In sum, regardless of the precise status of the landless farm household
heads not enumerated on schedule IV, it is clear that the supply of available
white labor was substantially greater in East Tennessee than in either of
the other two sections. Granted, these landless whites did not constitute a
perfectly stable and reliable labor force. Whereas "a slave who was bought,
stayed bought," landless whites were prone to move away or become farmowners in their own right.[38] Even so, it would be wrong to minimize their
contribution for at least three reasons. First, their numbers were consistently large throughout the 1850s. Second, as will be shown in the following
chapter, they were relatively stable geographically, and third, the majority
of those who remained in the same county throughout the 1850s failed to
acquire farms of their own. More than any other factor, their presence

36. Dyer and Moore, comps., *Tennessee Veterans Questionnaires*, vol. 4, pp. 1709, 1408.
37. Although they failed to explore the significance of their finding, the Owsleys did
    note their surprise in discovering that "a smaller proportion of nonslaveholders
    owned land in East Tennessee than in Middle or West Tennessee." Bode and
    Ginter also observed that in antebellum Georgia rates of landlessness were signif-
    icantly higher outside of the heavy cotton-growing regions. Indeed, they found that
    in the Georgia Black Belt the supply of white labor could vary markedly within a
    county, depending on soil characteristics as well as other factors. See Frank L. and
    Harriet C. Owsley, "The Economic Structure of Rural Tennessee, 1850–1860,"
    Journal of Southern History 8 (1942):162–82 (quotation on p. 165); Bode and Gin-
    ter, *Farm Tenancy and the Census*, pp. 136–44.
38. Wright, *The Political Economy of the Cotton South*, p. 49.

helps to explain the modest success of eastern landowners in overcoming the labor constraint imposed by the dearth of slave labor in their region.

Self-sufficiency and Commercial Orientation

Theoretically, slaveownership affected not only the scale on which slaveowners operated but also the focus of their operations, specifically, the relative balance between subsistence and commercial production on the individual farm. More often than not, contemporaries who commented on the subject were struck by the pervasive commercial orientation of antebellum farmers, in both North and South. Alexis de Tocqueville, for example, believed that most American farmers "made agriculture itself a trade." For generations historians agreed unquestioningly, assuming that the desire to produce for the market and to accumulate wealth had been nearly universal among nineteenth-century farmers. In the words of Richard Hofstadter, "the farmer himself, in most cases, was in fact inspired to make money, and such self-sufficiency as he actually had was usually *forced upon him* by a lack of transportation or markets."[39] Over the past two decades numerous scholars have rejected this entrepreneurial interpretation of American agricultural history. They argue that social and cultural factors militated against an unqualified commitment to market production. For many farmers, at least, profit maximization and wealth accumulation were "subordinate to . . . two other goals: the yearly subsistence and the long-run financial security of the family unit."[40]

Even when markets were accessible, then, many historians now question whether commitment to the market was automatic – except in the case of

39. Alexis de Tocqueville, *Democracy in America* (New York: Doubleday, 1969), p. 554; Richard Hofstadter, "The Myth of the Happy Yeoman," *American Heritage* 7 (1956):43, emphasis added.

40. The seminal essay in this regard is James A. Henretta, "Families and Farms: Mentalité in Pre-Industrial America," *William and Mary Quarterly* 35 (1978):3–32 (quotation on p. 19). See also Christopher Clark, "Household Economy, Market Exchange, and the Rise of Capitalism in the Connecticut Valley, 1800–1860," *Journal of Social History* 8 (1979):169–89. For works regarding the South specifically, see Steven Hahn, *The Roots of Southern Populism: Yeoman Farmers and the Transformation of the Georgia Upcountry, 1850–1890* (New York: Oxford University Press, 1983), pp. 29–39; John Thomas Schlotterbeck, "Plantation and Farm: Social and Economic Change in Orange and Greene Counties, Virginia, 1716 to 1860" (Ph.D. diss., The Johns Hopkins University, 1980), p. 22; and David F. Weiman, "Petty Commodity Production in the Cotton South: Upcountry Farmers in the Georgia Cotton Economy, 1840–1880" (Ph.D. diss., Stanford University, 1983), pp. 8ff.

farmers who owned slaves. "Simply put," as one scholar observes, "slavery *required* production for the market." Because slaves were expensive to begin with and were heavily taxed as well, those who purchased slaves implicitly committed themselves to at least enough market production to cover both purchase price and tax burden.[41] This heightened commercial focus not only affected the *mentalité* of producers and undermined "traditional modes of social interaction" within farm communities, as many social historians have argued. It also threatened more mundane consequences by reducing the South's capacity to provide its own foodstuffs. Since, in the Cotton South at least, production for the market implied concentration on a non-food staple, historians have long questioned the degree to which southern farmers were able to feed themselves and their region.[42]

Two observations are in order before turning to the Tennessee data. The first concerns the terms of the historical debate and their applicability to the Upper South. As with so many other important aspects of nineteenth-century southern history, scholars have focused primarily on the cotton-growing regions of the Deep South and have only rudimentary knowledge of patterns of food production elsewhere.[43] Efforts to assess the relative

41. Oakes, *Slavery and Freedom*, p. 97, emphasis added; Wright, *The Political Economy of the Cotton South*, pp. 7, 73–4.

42. Among the major contributions to this extensive debate, see Douglass C. North, *The Economic Growth of the United States, 1790–1860* (Englewood Cliffs, NJ: Prentice-Hall, Inc., 1961), Chap. 7; Albert Fishlow, "Antebellum Interregional Trade Reconsidered," in Ralph Andreano, ed., *New Views on American Economic Development* (Cambridge: Schenkman Publishing Company, 1965), pp. 187–200; Robert E. Gallman, "Self-Sufficiency in the Cotton Economy of the Antebellum South," *Agricultural History* 44 (1970):5–23; Raymond C. Battalio and John Kagel, "The Structure of Antebellum Southern Agriculture: South Carolina, A Case Study," *Agricultural History* 44 (1970):25–37; Dianne Lindstrom, "Southern Dependence upon Interregional Grain Supplies: A Review of the Trade Flows, 1840–1860," *Agricultural History* 44 (1970):101–13; William K. Hutchinson and Samuel H. Williamson, "The Self-Sufficiency of the Antebellum South: Estimates of the Food Supply," *Journal of Economic History* 31 (1971):591–612; Sam Bowers Hilliard, *Hog Meat and Hoecake: Food Supply in the Old South, 1840–1860* (Carbondale, IL: Southern Illinois University Press, 1972); and Robert E. Gallman and Ralph V. Anderson, "Slaves as Fixed Capital: Slave Labor and Southern Economic Development," *Journal of American History* 64 (1977):24–46. Most historians now accept that the antebellum South as a whole was generally self-sufficient in the production of food, although they are unsure whether the individual southern farmer typically aimed at self-sufficiency.

43. For exceptions to the pattern, see Schlotterbeck, "Plantation and Farm"; Walter Martin, "Agricultural Commercialism in the Nashville Basin, 1850–1860" (Ph.D. diss., University of Tennessee, 1984); Mary Beth Pudup, "The Limits of Subsistence: Agriculture and Industry in Central Appalachia," *Agricultural History* 64

Table 1.5 Household manufactures in Tennessee, 1840-1860

	East	Middle	West	State
Value per capita in 1840	$3.41	$3.74	$3.45	$3.58
Value per capita in 1860	$2.89	$2.65	$2.16	$2.57

Source: Derived from Rolla Milton Tryon, *Household Manufactures in the United States: A Study in Industrial History* (Chicago: University of Chicago Press, 1917), pp. 334-7.
Note: Figures are based on aggregate data from all Tennessee counties. Apparently defective data for Roane and Wilson counties in 1860 have been treated as missing.

strengths of subsistence vs. commercially oriented motives have rested on the assumption that self-sufficiency and commercial production were competing goals and have devolved into intensive analyses of the crop mix on individual farms (i.e., the relative emphases on cotton and corn). Although this approach makes sense for the Cotton South (including the sample counties in West Tennessee), it offers little insight when applied to major food-producing regions such as Middle and East Tennessee, areas where food production and profit maximization were mutually reinforcing strategies.

Second, it is important to note the very narrow definition of self-sufficiency that scholars have adopted in assessing the production patterns of antebellum farmers, one that focuses exclusively on household needs for grain and meat. By the mid-nineteenth century there may have been occasional isolated communities that were, in the broadest sense of the term, self-sufficient. However, if we mean by the term the ability to provide not only basic foodstuffs but also farm implements, clothing, medicinal supplies, and so on, then by the late antebellum period the truly self-sufficient household was by definition extraordinarily rare.[44]

This generalization held as true in Tennessee as elsewhere. A good indicator of the dearth of a more inclusive self-sufficiency is the minimal quantity of household manufactures that the state produced by the mid-nineteenth century (see Table 1.5). For the South as a whole the value of

(1990):61–89; idem., "The Boundaries of Class in Preindustrial Appalachia," *Journal of Historical Geography* 15 (1989):139–62.
44. Ford, *Origins of Southern Radicalism*, pp. 81–4; Rodney C. Loehr, "Self-Sufficiency on the Farm," *Agricultural History* 25 (1952):37–41.

household manufactures per capita had dropped to only $1.60 by 1860. The per-capita figure was almost a full dollar higher in Tennessee for the same year, but even that level of production was strikingly low. By comparison, economic historians have estimated that southern slaves in 1860 consumed manufactured items valued at between $7.70 and $11.00 per capita; consumption levels should have been substantially higher among southern whites.[45] Clearly, the vast majority of the state's farmers were engaged in some degree of commercial activity. Although there were slight interregional variations in the importance of household manufactures across the state, farmers in all sections of the state were involved at least nominally in the market economy.

Employing census data regarding livestock inventories and crop outputs, and relying on contemporary estimates concerning the consumption requirements of animals and humans, it is relatively simple to determine the degree to which Tennessee farmers satisfied the more narrow definition of self-sufficiency – that is, adequate production of grain and meat. Total food production can be estimated by summing the yield of grains, peas and beans, and potatoes (reduced to reflect seed allowances for the following year and converted to corn-equivalent units for comparability) and adding to that total the estimated output of beef, mutton, and pork (inferred from livestock inventories and grain surpluses and also converted to corn-equivalent units at the ratio of 7.6 net pounds of meat per bushel of corn). An index of self-sufficiency in total foodstuffs can then be constructed, equivalent to the ratio of total food production to the estimated quantity of food necessary for the subsistence of both humans and livestock. An index above 1.0 would indicate some measure of surplus food production, whereas an index below 1.0 would imply a food deficit.[46]

As a whole, the state of Tennessee at the close of the antebellum era was abundantly self-sufficient in foodstuffs (see Table 1.6, figures for "Entire population"). As expected, the index was lowest for the sample counties of West Tennessee, where commercially minded farmers devoted much of their energy and acreage to a nonfood crop. The West Tennessee index of 1.14 indicates that the district's sampled farmers produced quantities of

45. Roger Ransom and Richard Sutch, *One Kind of Freedom: The Economic Consequences of Emancipation* (Cambridge: Cambridge University Press, 1977), p. 211. The lower- and upper-bound estimates include manufactured items produced on the farm as well as those purchased in the marketplace.
46. For a discussion of the self-sufficiency index, see Hilliard, *Hog Meat and Hoecake*, p. 158. For a full explanation of the methodology employed in the self-sufficiency analysis, see Appendix B.

Table 1.6 Indexes of self-sufficiency in foodstuffs and proportion of farm surplus consumed locally, sample Tennessee counties, 1859

	East	Middle	West
Entire population[a]	1.37	1.38	1.14
Farm operators only			
All operators	1.80	1.51	1.24
Owner-operators	1.84	1.53	1.23
Tenants	1.55	1.44	1.27
Estimated proportion of food surplus consumed locally:	43.0%	27.8%	42.4%

[a]Includes entire farm and nonfarm population.
Source: Eight-county sample, see text.

grain and meat that exceeded total combined human and animal requirements by a minimum of 14 percent.[47] It was substantially higher and virtually identical in the mixed-farming sample counties of Middle and East Tennessee, 1.38 and 1.37, respectively. Given the extensive differences among the regions already discussed, however – in particular the disparity in importance of slavery and in scale of farm operations – distinctions among the regions in overall food production seem surprisingly small.[48]

The examination of household-level data from the eight sample counties can reveal far more, however, than a cursory evaluation of statewide patterns. Because the regional indexes given in row one of Table 1.6 account for food consumed by the local nonproducing population (nonfarm households as well as households headed by farmers and farm laborers without

47. Actual aggregate food production ought to have been even higher than these figures suggest, because if enumerators followed instructions, they excluded all farms with produce valued at less than $100. My thanks to Fred Bode for this important point.

48. The eight sample counties, although not scientifically selected or statistically representative of the entire state in a mathematically verifiable sense, nonetheless closely mirrored the larger regions from which they were drawn in patterns of food production, as analysis of aggregate published data for all eighty-four of the state's counties shows. The overall index of self-sufficiency in total foodstuffs was 1.39 for the thirty-three East Tennessee counties, 1.31 for the thirty-four Middle Tennessee counties, and 1.18 for the eighteen counties in West Tennessee. The index of self-sufficiency for the entire state in 1859, including the nonfarm population, was 1.30.

farms), they underestimate the surplus production of local farm operators. For the sample counties in each region, therefore, the indexes of self-sufficiency for farm operators only are higher than for the entire regional population. Overall, farm operators in all three regions produced far more grain and meat in 1859 than their own households required. In Middle Tennessee, for example, operators produced roughly 1.5 times their basic household needs, and farmers in East Tennessee almost doubled their household demands. Not unexpectedly, these findings clearly endorse the reputation of central and eastern Tennessee as among the leading food-producing regions of the Old South. More surprising is the level of food production among West Tennessee farm operators. Farmers in Fayette and Haywood counties were heavily committed to commercial cotton produc-tion – they grew nearly 62,000 bales of cotton in 1859 – yet they still produced enough corn and pork to exceed their own requirements by one-fourth.

Clearly, farmers in every region produced a surplus of foodstuffs that was consumed off the farm. It is less clear, however, whether this reflected a direct commitment to market production. When Fayette County farmers grew cotton or Robertson County farmers grew tobacco, they obviously did so with the intention of selling their output in the marketplace. Resources used to grow those crops had to be diverted from resources used to grow food for the household. The commercial production of foodstuffs, on the other hand, was simply an extension of production for household needs. The farmer who had foodstuffs left over at the end of the year may have intended from the outset to produce a surplus for market exchange. On the other hand, he may have followed a "safety-first" strategy, and pro-duced enough food to ensure self-sufficiency even under the worst possible weather conditions. In normal years, then, farmers who adopted this strat-egy would have held a marketable surplus at year's end.[49]

In all three regions, however, farm operators must have decided in ad-vance to produce foodstuffs for sale or exchange in the market.[50] With the

49. For the concept of safety-first operations, see Gavin Wright and Howard Kun-reuther, "Cotton, Corn, and Risk in the Nineteenth Century," *Journal of Economic History* 35 (1975):526–51.

50. There is no evidence to suggest that the magnitude of the surplus was unexpected (i.e., that the 1859 grain crop was exceptionally good). Census enumerators in each of the sample counties rated crop outputs for 1859 as fair or somewhat below average. See U.S. Census Office, 8th Census [1860], Schedule VI, Social Statistics (manuscript schedules). For evidence that the 1859 corn crop in the Cotton South was below average, see Donald F. Schaefer, "The Effect of the 1859 Crop Year

possible exception of West Tennessee, the magnitude of the surplus is simply too great to constitute only the margin of safety planned by risk-averse farmers. In every region a large proportion of farmers evidently began the 1859 planting season with the intention of selling foodstuffs for market exchange. The intention of selling surplus foodstuffs was not necessarily equivalent to a commitment to external markets, however. There was evidently an active local market for foodstuffs in every region, particularly in the eastern and western sample counties where more than two-fifths of the farm surplus could have been consumed locally, as Table 1.6 shows. Even so, in all three sections more than one-half of local food surpluses must have been traded externally.

The actual destination of exported foodstuffs is impossible to determine. The magnitude of the western food surplus relative to local needs is too small to establish conclusively that those counties were exporting foodstuffs. Part of the surplus may simply reflect sampling error; part may have been absorbed by a higher level of consumption or by spoilage. Each of the sample western counties was connected to Memphis by either water or rail, however, and they may have sent some surplus foodstuffs to supply that inland commercial center.

Food surpluses in Middle and East Tennessee were proportionally much larger than to the west, and the disposition of the surpluses is much less easily explained. Part of the food surplus in Middle Tennessee may have also been sent to meet urban demand in Nashville, but that city in 1860 was not large enough to absorb more than a tiny fraction.[51] Urban demand was even less significant in East Tennessee. Knoxville, the region's largest "city," contained fewer than 4,000 people in 1860. That farmers continued to produce large food surpluses despite an adequate urban market within Tennessee suggests that a demand for their foodstuffs existed outside the state.

Some of the surplus undoubtedly was sent northward. In 1856 alone 100,000 bushels of Tennessee wheat were sold in New York, and observers estimated that annual demand would soon exceed 300,000 bushels. In all likelihood, however, most of the surplus was probably exported to the Lower South in the form of beef and pork, the latter especially, either in

Upon Relative Productivity in the Antebellum Cotton South," *Journal of Economic History* 43 (1983):859–60.

51. The population of Nashville in 1860 was approximately 17,000. Assuming that none of the other surrounding counties in the region sent foodstuffs into the city, the surplus from the three Middle Tennessee sample counties alone was sufficient to feed the city for over two and one-half years.

the barrel or on the hoof. Smith County farmer James Young likely spoke for most Middle Tennesseans when he tersely informed a government agent that "we sell considerable pork." Farmers from the region could drive surplus cattle and hogs directly overland into the cotton-growing regions of Alabama and western Georgia or send their herds via river or rail to Nashville, which was emerging as an important meatpacking center. The processed meat could then be shipped eastward on the Nashville and Chattanooga Railroad and then, via rail connections accessible from the latter city, routed on to Charleston, Savannah, or the interior of the South Atlantic states.[52]

Likewise, East Tennessee farmers drove cattle and hogs by the tens of thousands into the Carolinas or floated them down the Tennessee into the Lower South. *Debow's Review* estimated that in 1849–50 alone 81,000 hogs were driven to the East Coast from the mountains of Tennessee and Kentucky. Charles Lanman, a New Yorker who visited the region in 1848, noted that "the principal revenue of the people . . . is derived from the business of raising cattle, which is practiced to a considerable extent. The mountain ranges afford an abundance of the sweetest grazing food, and all that the farmer has to do in the autumn is to hunt up his stock, which has now become excessively fat, and drive them to the Charleston or Baltimore market." This "sequestered part of the Union" was without a major railroad until almost the end of the antebellum era, but after 1858, with the completion of the East Tennessee and Virginia Railroad (which connected the region with Chattanooga and the Deep South), eastern farmers could also export their surplus by rail.[53]

52. See Clark, *The Tennessee Yeomen*, p. 118; United States Patent Office, *Annual Report of the Commissioner of Patents for the Year 1848* (Washington, DC: Wendell & Van Benthuysen, 1849), pp. 525, 505, 516. According to another correspondent of the Commissioner of Patents during the same year, "a large proportion of the planters in South Alabama . . . [had] been dependent on Tennessee for their pork." Also, a correspondent from Louisiana noted that planters in that state had "been compelled to buy their corn from Kentucky and Tennessee." See also Gray, *History of Agriculture in the Southern United States*, pp. 836–41; Clark, *The Tennessee Yeomen*, pp. 121–8; Stephen V. Ash, *Middle Tennessee Society Transformed, 1860–1870: War and Peace in the Upper South* (Baton Rouge: Louisiana State University Press, 1988), pp. 16–17. On Middle Tennessee rail connections see Killebrew, *Introduction to the Resources of Tennessee*, pp. 314ff. For general arguments that the Upper South sent large quantities of foodstuffs into the Lower South, especially the Gulf and Atlantic coasts, see Hilliard, *Hogmeat and Hoecake*, pp. 192–6, and Lindstrom, "Southern Dependence Upon Interregional Grain Supplies."
53. Charles Lanman, *Letters from the Allegheny Mountains* (New York, 1848), p. 153. See also United States Patent Office, *Annual Report of the Commissioner of Patents,*

Table 1.7 Median marketable proportion of total farm production, sample Tennessee counties, 1859 (percentages)

	East	Middle	West
All operators	40.8	37.0	67.6
By tenure			
Owner-operators	41.6	37.7	67.8
Tenants	39.2	33.0	64.7
By farm size			
1-24 improved acres	2.8	26.6	50.1
25-49 improved acres	26.4	38.1	55.7
50-99 improved acres	32.2	39.8	67.1
100-199 improved acres	45.4	40.0	69.4
200+ improved acres	54.5	40.6	72.1

Source: Eight-county sample, see text.
Note: Marketable production is defined to equal the wholesale value of all nonfood crops plus the value of foodstuffs produced above the quantity required for household and live-stock consumption. The marketable proportion of total farm production represents the ratio of the value of marketable production to that of total production, defined as the esti-mated value of household manufactures plus the wholesale market value of all crop and livestock output produced on the farm, whether consumed on the farm or traded in the marketplace. (See Appendix C for wholesale price data.)

Because farmers in eastern and central Tennessee were so heavily in-volved in the commercial production of foodstuffs, the differences among the regions in degree of commercial orientation are not as great as one would expect when comparing a region characterized by plantations and cotton fields with areas denoted by small-scale, mixed farming. Table 1.7 presents estimates for Tennessee farms of the average ratio of the value of the marketable surplus – that is, the value of goods not consumed on the farm and thus available for market exchange – to the value of total pro-duction, a statistic that varied with the extent of market involvement. More than two-thirds of the total output of West Tennessee farmers was traded in the marketplace, a figure that bears powerful witness to their pronounced commercial orientation. In truth, however, the average proportion sold off the farm was impressively high in all three regions of the state. If farmers

p. 521; Edmund Cody Burnett, "Hog Raising and Hog Driving in the Region of the French Broad River," *Agricultural History* 20 (1946):87–103.

in eastern and central Tennessee were relatively less involved in market exchange, they still sold roughly two-fifths of their farm output, hardly an indication of market estrangement.

Certainly it is the case that the nature of market involvement and the typical direction of market ties differed among the regions, with western cotton producers more likely than farmers elsewhere to be enmeshed in larger economic networks, producing for national or international rather than local markets.[54] These were differences of degree rather than of kind, however, a fact that cannot be too highly stressed. In West Tennessee, where cotton was "King," planters not only produced the white fiber for the factories of England but also participated in a brisk local market for corn and pork. In Middle and East Tennessee, on the other hand, farmers with far smaller operations nonetheless produced extensively for distant consumers.

Within each region owners and tenants were almost identical in degree of market emphasis. Both appear to have responded to market incentives in a highly similar fashion, although, as will be noted, a large fraction of tenants' surplus went to pay rent. In contrast to tenure form, the size of the farm unit evidently affected the extent of market involvement substantially, especially in eastern and western Tennessee (see Table 1.7). In Middle Tennessee farm size appears to have exercised little influence on degree of commercial involvement, except on the very smallest farms. In the other two sections, on the other hand, market involvement increased linearly with farm size. The relationship was most striking in East Tennessee. On the average, East Tennessee farmers with fewer than twenty-five improved acres sold only a tiny fraction (less than 3 percent) of their output off of the farm. At the other end of the farm size spectrum, farmers of 200 or more improved acres were selling more than half of their output. At the western end of the state, even the smallest farmers traded a majority of their output in the marketplace; among the largest cotton planters of the region the proportion increased to nearly three-fourths.

Although, in the aggregate, farm operators in all three sections maintained self-sufficiency while producing large surpluses for commercial exchange, it does not follow necessarily that each *individual* farmer was self-sufficient or contributed to this surplus. Even among those who focused exclusively on food production, the unpredictable caprice of nature – excessive rainfall, premature frosts, rust in the wheat, and so on – fre-

54. Schlotterbeck, "Plantation and Farm," pp. 1–9, 211–13.

quently frustrated farmers in their efforts to feed their own families, let alone produce a marketable surplus. In addition to such random variation, however, there also may have been systematic differences among farmers. Tenants, for example, may have consistently chosen production patterns different from those of owners. There is considerable evidence that in the postbellum South tenants were more frequently deficient in foodstuffs than were owners.[55] A comparable situation may have also obtained during antebellum years. Similarly, the size of the farm operation may have exercised an important influence on operators' crop choices and extent of commercial production.

To determine the proportion of individual households that were self-sufficient in foodstuffs, it may be desirable to modify slightly the method employed to estimate overall regional self-sufficiency in order to take into account the special circumstances faced by tenant farmers. Our knowledge of antebellum rental agreements is woefully inadequate, not least because until only recently scholars dismissed agricultural tenancy as an insignificant institution in the Old South.[56] Scattered contemporary references indicate that the rent paid by tenants took a variety of forms, most commonly a fixed amount of cash, a fixed quantity of the crops produced on the rental unit (standing rent), or a fixed proportion thereof, typically one-third of the grain and one-fourth of the cotton (the so-called thirds and fourths system). All other things the same, in a perfectly competitive market rents should have been approximately equal across the forms of tenure. In actuality, share tenants (i.e., renters who paid a share of their output to the landlord) likely paid somewhat higher rents than other types of tenants, the differential representing a premium paid to the landlord for sharing a portion of the risk associated with fluctuating crop yields and crop prices.[57] Table 1.8 presents two estimates of the incidence of self-sufficiency among individual households. The first, which makes no allowance for rental payments by tenants, is most appropriate for comparison with other studies because scholars have not typically made allowances for rental payments of

55. See, for example, Ransom and Sutch, *One Kind of Freedom*, pp. 156–65.

56. Gray, *History of Agriculture in the Southern United States to 1860*, pp. 646–67; Ransom and Sutch, *One Kind of Freedom*, p. 88. The most convincing rebuttal is Bode and Ginter, *Farm Tenancy and the Census*, but see also Joseph D. Reid Jr., "Antebellum Southern Rental Contracts," *Explorations in Economic History* 13 (1976):69–83.

57. Joseph D. Reid Jr., "Sharecropping as an Understandable Market Response: The Post-Bellum South," *Journal of Economic History* 33 (1973):113; Robert Higgs, *Competition and Coercion: Blacks in the American Economy, 1865–1914* (Chicago: University of Chicago Press, 1977), pp. 50ff.

Table 1.8 Percentage of farm operators self-sufficient in food production, sample Tennessee counties, 1859

	East	Middle	West
	Grain		
All operators	84.3 (81.6)	82.4 (79.0)	74.1 (70.3)
Owner-operators	84.5	83.9	74.4
Tenants	83.5 (67.2)	76.7 (60.1)	72.2 (44.6)
	Meat		
All operators	82.9 (80.1)	80.7 (77.4)	75.1 (72.4)
Owner-operators	84.2	84.5	76.1
Tenants	76.5 (60.0)	65.8 (49.6)	68.9 (49.1)
	Total foodstuffs		
All operators	79.7 (76.6)	78.2 (74.7)	68.9 (65.1)
Owner-operators	80.9	82.0	69.7
Tenants	73.7 (55.3)	63.1 (45.8)	63.5 (36.3)

Source: Eight-county sample, see text.
Note: Figures in parentheses are adjusted rates of self-sufficiency after tenant grain production is reduced by one-third to reflect rent payments.

grain in evaluating the self-sufficiency of antebellum farmers.[58] The second estimate (given in parentheses), computed after reducing tenants' grain production by one-third to reflect the payment of rent, should be interpreted as a definite lower bound. Because the census data employed in the analysis do not include a whole range of food sources – garden products, poultry, fish, and wild game, for example – the rent-adjusted estimate almost certainly understates the true extent of self-sufficiency on Tennessee farms.

58. For an analysis of postbellum self-sufficiency that makes such allowances, see Ransom and Sutch, *One Kind of Freedom*, pp. 151–9, 251–3.

Although the sampled farm communities in all three regions of the state were lavishly self-sufficient in the production of food, in all three sections numerous individual operators were unable to feed their families. Not surprisingly, the proportion achieving self-sufficiency in total foodstuffs was lowest in West Tennessee, where farmers devoted much of their acreage to King Cotton. At a minimum, nearly one-third (31 percent) of farmers in that region produced less grain and meat than their own households required. Less expected is the relatively high proportion of deficient households among farm operators in Middle and East Tennessee, where 20 to 25 percent of farmers failed to achieve self-sufficiency, despite the fact that family subsistence and market production were complementary rather than competing goals. The Central Basin and the Valley of East Tennessee ranked with the Shenandoah Valley of Virginia and the Bluegrass Region of Kentucky as the leading food-producing regions of the Old South, yet at least one-fifth of farm operators in those areas could not feed their own families from the produce of their own farms.

Although interregional comparisons within the state are straightforward, it is difficult to interpret the various estimates in an absolute sense. In other words, were the rates of self-sufficiency exhibited by the sample Tennessee counties high or low when evaluated within a larger regional or national context? In this regard the meticulous analysis of more than 11,000 antebellum northern farms by Jeremy Atack and Fred Bateman provides a convenient and interpretively useful benchmark of comparison for the Tennessee data. Assessing the "attainments of the yeoman agricultural system at the peak of its prominence," Atack and Bateman found that in 1859 37 percent of northeastern farms and 25 percent of midwestern farms were deficient in the production of food and feed; for the rural North as a whole the proportion of farms with overall food deficits was 29 percent.[59]

In light of these figures, the rates of self-sufficiency among Middle and East Tennessee farmers were indeed high; they equaled or exceeded those for leading midwestern states such as Indiana, Minnesota, and Wisconsin. Furthermore, with northern patterns in mind, it would seem inappropriate to label the rate of self-sufficiency in the cotton counties of West Tennessee as low; although substantially below that for other sections of the state, it

---

59. Atack and Bateman, *To Their Own Soil*, p. 223; the quotation is from Atack and Bateman, "Yeoman Farming: Antebellum America's Other 'Peculiar Institution,' " in Lou Ferleger, ed., *Agriculture and National Development: Views on the Nineteenth Century* (Ames: Iowa State University Press, 1990), p. 26. By comparison, Steven Hahn estimates that 74 percent of farmers in the Georgia upcountry were self-sufficient in 1850. See Hahn, *The Roots of Southern Populism*, p. 32.

nonetheless exceeded rates for major food-producing states such as Penn-
sylvania, Ohio, and Michigan. In the pervasiveness of self-sufficiency, if
not the size of the total marketable surplus, *all* regions of the state – regardless of degree of dependence on slavery or commitment to nonfood
staples – compared favorably with the rural North.[60]

When the non-farm-operating sector of the Tennessee counties is also
considered (farmers and farm laborers without farms, plus nonagricultural
households), it is abundantly apparent how crucial the local exchange of
foodstuffs was, even in major food-producing areas. Considering all households in each region, the maximum possible proportion able to meet subsistence needs independently was only 44.1 percent in West Tennessee,
55.3 percent in the central counties, and 44.3 percent in the eastern section.
Although community self-sufficiency characterized all three sections, a clear
majority of households in each rural area were either buying or selling
foodstuffs in local markets.[61]

Although numerous factors undoubtedly influenced the propensity
among farm operators for household self-sufficiency, two factors appear to
have been crucial. The first was the tenure status of the operator. Despite
the great attention that scholars have paid to the farm operations of tenants
after the Civil War, they have very nearly ignored the influence of tenure
on antebellum farming. Before adjustments are made to reflect rental payments, that influence appears to have been relatively small. As Table 1.8
indicates, tenants were more likely to require additional foodstuffs than
were farmers who owned and operated their own farms. This was despite
the fact that, overall, tenants like owner-operators produced far more grain
and meat than their families and livestock required. As Table 1.6 shows,
tenants produced between 27 and 55 percent more total foodstuffs than

60. The methodology that Atack and Bateman employ to determine self-sufficiency is
    not identical to that used in this study. In particular, they also incorporate estimates
    of the consumption of dairy products on northern farms. Although this constitutes
    a more demanding standard of self-sufficiency than that used here, the feed re-
    quirements for work stock that I employ are substantially higher than theirs, a
    factor that should tend to offset this disparity. See *To Their Own Soil*, pp. 208–
    14, and Appendix B of this work.
61. It is worth remembering that urban demand had little to do with this impressive
    degree of local trade; there was not a single town in the sample counties with as
    many as 1,000 inhabitants. On the importance of local exchange networks, see
    Hahn, *The Roots of Southern Populism*, p. 39; John T. Schlotterbeck, "The 'Social
    Economy' of an Upper South Community: Orange and Greene Counties, Virginia,
    1815–1860," in Orville Vernon Burton and Robert C. McMath Jr., *Class, Conflict,
    and Consensus: Antebellum Southern Community Studies* (Westport, CT: Greenwood
    Press, 1982), pp. 20–5.

household demands necessitated. In both Middle and West Tennessee the level of food production was highly similar between tenants and owner-operators, with tenants actually exceeding owners in the latter region in the proportional production of food. Only in East Tennessee did there exist a sizable gap in food indexes between renters and owners, but the difference in that section was less a reflection of low production among tenants than of extraordinarily high levels of food production among owner-operators.

Before allowances are made for rent payments, rates of self-sufficiency in grain production among tenants are roughly comparable to those of owner-operators. By far the largest difference existed in Middle Tennessee, where the proportion of operators producing sufficient quantities of grain was about 7 percentage points higher among owners than among tenants. In all three sections the tenure-associated difference was considerably larger with regard to meat production. Because tenants generally had less financial capital than did owners, they perhaps were unable to invest as much proportionately in livestock. For farmers with access to the open range, however, the expense of feeding swine was relatively slight until the last two months or so before slaughter, when Tennessee farmers commonly pen-fed their hogs. Perhaps the lower level of meat production simply represented the present-mindedness of some tenants. Meat production took considerably longer to provide a return on investment than did grain production. (It generally took at least a year and a half to fatten hogs for slaughter, and as much as twice as long to fatten cattle.) Consequently tenants, who generally operated on one-year leases, may have preferred to maximize short-run returns by devoting relatively fewer resources to the production of meat.[62] All things considered, however, differences across tenures were still not large – allowances for sampling error could erase the gap entirely in East and West Tennessee – and the typical tenant in every section clearly produced sufficient meat for his family.

Reducing the grain production of tenants to reflect rent payments slightly lowers overall self-sufficiency rates but drastically reduces estimates of self-sufficiency among tenants specifically and dramatically alters our

---

62. Forrest McDonald and Grady McWhiney, "The Antebellum Southern Herdsmen: A Reinterpretation," *Journal of Southern History* 41 (1975): 147–76; McDonald and McWhiney, "The South from Self-Sufficiency to Peonage: An Interpretation," *American Historical Review* 85 (1980): 1095–118; Donald L. Winters, *Farmers Without Farms: Agricultural Tenancy in Nineteenth-Century Iowa* (Westport, CT: Greenwood Press, 1978), p. 40. On swine-feeding practices among Tennessee farmers, see Hutchinson and Williamson, "The Self-Sufficiency of the Antebellum South," p. 600.

perception of tenant operations. The maximum proportion of tenants deficient in grain production after the rent was paid ranged from approximately one-third in East Tennessee to over one-half in West Tennessee. The reduction of grain supplies affected potential meat output as well, for a primary use of surplus grain was as a feed supplement for livestock being fattened for slaughter. When allowances for reduced animal feed are made, the maximum proportion of tenants producing inadequate supplies of meat rises to between two-fifths and one-half. When both categories are combined to consider total foodstuffs, the proportion of deficient tenant households increases to approximately one-half in Middle and East Tennessee and to nearly two-thirds in West Tennessee. Even though these estimates are upper bounds, it is apparent that long before the disruption of civil war and emancipation, long before the collapse of the southern credit structure and the rise of the insidious crop lien, Tennessee tenants found it a difficult challenge both to feed themselves *and* to pay the rent.

A second factor that clearly influenced the propensity for basic self-sufficiency was the size of the operator's farm (see Table 1.9). In order not to exaggerate the independent influence of farm size, Table 1.9 presents self-sufficiency rates by farm size both before and after adjustments for rental payments by tenants. In every region the likelihood of self-sufficiency tended to increase as farms grew larger. This relationship was most pronounced in East Tennessee. A majority of the smallest farms in that section (those under twenty-five improved acres) were deficient in total foodstuffs, whereas self-sufficiency in the largest category of farms (200 or more improved acres) was nearly universal. In Middle and West Tennessee the propensity for self-sufficiency declined slightly for the largest category of farm units, but it was still the case in those regions that farms of more than 200 acres were more likely to be self-sufficient in foodstuffs than were farms of under 100 acres.

The figures for West Tennessee bolster the contention that large-scale cotton production was logically consistent with self-sufficiency in food.[63] Evidently very few small farmers in the region were selling surplus foodstuffs to large cotton planters; if anything, it is more likely that planters were supplying small farmers. This must have been the case with West

---

63. Gallman and Anderson, "Slaves as Fixed Capital." For competing views that question the complementarity of cotton and corn production, see Weiman, "Petty Commodity Production in the Cotton South," pp. 126, 213; Hilliard, *Hog Meat and Hoecake*, p. 152.

Table 1.9 Food production patterns by size of farm, sample Tennessee counties, 1859

	East	Middle	West
Index of self-sufficiency in total foodstuffs, for farms with:			
1-24 improved acres	0.94	1.20	1.19
25-49 improved acres	1.25	1.41	1.18
50-99 improved acres	1.47	1.57	1.26
100-199 improved acres	1.85	1.67	1.37
200+ improved acres	2.36	1.41	1.22
Proportion self-sufficient in total foodstuffs, for farms with:			
1-24 improved acres	46.7 (42.8)	60.2 (52.8)	50.4 (40.3)
25-49 improved acres	62.8 (60.2)	79.4 (72.6)	64.2 (54.6)
50-99 improved acres	77.7 (76.5)	84.2 (83.8)	64.4 (60.5)
100-199 improved acres	91.2 (90.3)	93.3 (92.7)	80.5 (78.4)
200+ improved acres	97.2 (97.2)	88.7 (88.7)	75.4 (74.8)

Source: Eight-county sample, see text.
Note: Figures in parentheses are adjusted rates of self-sufficiency after tenant grain production is reduced by one-third to reflect rent payments.

Tennessean John W. Jones, for example. The largest landholder and slaveholder in Fayette County, Jones grew 800 bales of cotton in 1859, worth easily in excess of $32,000. Significantly, he also raised 600 swine and grew 12,000 bushels of corn, producing in all almost 60 percent more food than necessary to provide amply for his work stock, his family, and his 180 slaves. Jones's surplus, probably traded locally, was sufficient to feed thirty slaveless families. On the other hand, the West Tennessee data would seem to contradict the argument that small farmers in the antebellum Cotton South followed a safety-first strategy by planting more than enough food

Table 1.10 The crop mix (pounds of cotton per bushel of corn) on sample West Tennessee farms by size of farm, 1859

1-24	improved acres	7.8
25-49	improved acres	9.6
50-99	improved acres	14.5
100-199	improved acres	15.6
200+	improved acres	23.8

Source: Sample of farms from Fayette and Haywood counties, see text.

crops to ensure family self-sufficiency and treating cotton as a secondary crop only.[64] At first glance, at least, it would appear that small farm operators in the western counties were frequently quite willing to forgo self-sufficiency for the potential profits of cotton.

In actuality, a more accurate characterization of the behavior of West Tennessee smallholders would lie somewhere between these two extremes. As Table 1.10 shows, emphasis on cotton production in the western sample counties did in fact increase dramatically with farm size. The cotton/corn ratio on farms under twenty-five improved acres was two-thirds lower than on plantations of 200 acres or more. As the safety-first hypothesis would indicate, small farmers in West Tennessee were indeed more cautious than large planters, evidently devoting greater proportions of their acreage to food crops and relatively smaller proportions to the more risky nonfood staple. In West Tennessee, however, it was simply not the case that the typical small farmer was *sufficiently cautious* to achieve household self-sufficiency.[65] Even before adjusting for the rent payments of tenants (most of whom operated small farms), more than two-fifths of farmers with fewer than fifty improved acres were deficient in basic foodstuffs.

Gross Farm Income in 1859

Although the majority of farmers in every size and tenure category produced more than enough to feed themselves, the total value of farm pro-

64. Wright, *The Political Economy of the Cotton South*, pp. 62–74.
65. In his discussion of self-sufficiency and risk, Gavin Wright implies without proving that individual farmers were typically self-sufficient. See *The Political Economy of the Cotton South*, pp. 62–74.

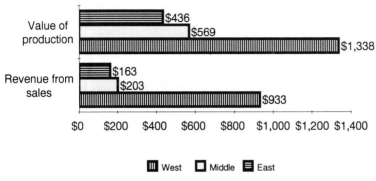

Figure 1.3 Median value of production and revenue from sale of farm products, sample Tennessee counties, 1859

duction, the gross revenue from sales of farm products, and the net income from farm operations varied widely, not only among but also within each region of the state.

Antebellum censuses did not ask farmers to estimate the total value of their production, but by applying wholesale price data it is possible to produce a rough estimate thereof, equal to the total value of grain production, meat production, and household manufactures per farm.[66] As expected, the average value of total production per farm was much higher in West Tennessee than elsewhere in the state (see Figure 1.3). This was partially, but not primarily, a function of the larger size of farms in that region. As shown earlier, median farm size in West Tennessee was only 33 percent greater than in Middle Tennessee and 43 percent greater than in East Tennessee. The median value of total production per western farm, however, more than doubled that for Middle Tennessee farms and tripled that for eastern farms.

More than anything else, the much greater value of production on western farms reflected the concentrated focus of the region's farmers on cotton, a far more valuable crop per acre than either corn or wheat, the primary alternatives available to farmers in other areas. Because antebellum censuses did not record the amounts of acreage devoted to specific crops, it is impossible to determine crop yields precisely or to calculate the typical value of output per acre planted in cotton, corn, or wheat. Government statistics for the years 1866–1900, however, indicate that in Tennessee during that

66. An exact evaluation of gross farm income would also include the value of forest, garden, orchard, and dairy products and, where appropriate, income from rent. For the price data employed in these estimates, see Appendix C.

thirty-five-year period the value of output per acre planted in cotton was approximately double that per acre planted in corn. Because during the postbellum period as a whole the ratio of cotton to corn prices was lower than during the 1850s, the differential in 1859 was likely even greater than the postwar figures would suggest.[67]

The difference among the sections in average *revenue* from farm production (i.e., the value of output traded in the marketplace) is even greater than that for the value of total production, which included the value of products consumed on the farm as well as sold.[68] As Figure 1.3 shows, average revenue from sales per farm was more than four times greater in West Tennessee than in Middle Tennessee and nearly six times greater than in East Tennessee. The estimates in Figure 1.3, it should be noted, represent upper bounds for market-generated revenue. Undoubtedly, less than 100 percent of surplus food production was actually sold off the farm. Some was absorbed through higher levels of consumption, some merely wasted. Although the estimates consequently overstate farm revenue to an unknown degree, they still constitute approximately accurate indicators of relative differences among regions and between tenure categories.

The larger scale of farm operations in West Tennessee and the relatively greater degree of market involvement among the region's farmers – reflected partially through the sale of surplus foodstuffs but primarily in their emphasis on cotton, a more valuable crop per acre – led to a higher per-capita income and, potentially at least, to a much higher standard of living. The census does not provide enough information to construct an exact evaluation of farm income. Table 1.11 presents crude estimates of net per-capita farm income based on the values of total farm production depicted in Figure 1.3. To arrive at a rough approximation of farm income, the value of total production is modified to account for crops fed to livestock or retained for seed. The production of tenants is further reduced to reflect rent payments, and that of slaveowning operators is reduced to account for the costs of slave upkeep, including not only the value of food consumed by slaves but also estimated costs for clothing and medical sup-

67. Wright, *Old South, New South*, p. 36. The value of total production per improved acre was $6.59 in East Tennessee, $9.02 in Middle Tennessee, and $13.92 in West Tennessee. The value of production per capita (slave and free) was $70.71, $81.14, and $114.20 respectively.

68. I define revenue from sale of farm products to equal the wholesale value of all nonfood crops plus the value of foodstuffs produced above the quantity required for seed, livestock feed, and household consumption. (See Appendix C for wholesale price data.)

Table 1.11 Median net farm income per capita, sample Tennessee counties, 1859

	East	Middle	West
All operators	$40	$48	$152
Owner-operators	46	58	193
Tenants	19	20	51
Slaveholders	67	69	239
Slaveless	38	38	58
Market-oriented[a]	89	117	214
Subsistence-oriented[b]	27	34	22

[a]Able to sell one-half or more (by value) of total farm production off the farm after meeting subsistence needs. [b]Unable to sell as much as one-half (by value) of total farm production off the farm after meeting subsistence needs.
Source: Eight-county sample, see text.
Note: Monetary values are rounded up to the nearest dollar. For construction of income estimates, see text.

plies.[69] A more precise estimate would also take into consideration other sources of revenue, such as capital gains due to the physical improvement of the farm or the natural increase of livestock inventories, as well as additional costs such as machinery and implement depreciation or wages paid to hired laborers.[70]

The per-capita income estimates bring to mind Robert Fogel and Stanley Engerman's observation that, "however heavy the yoke of slavery, it produced considerable prosperity for free men."[71] Per-capita income on West

69. The value of total production is first reduced by the estimated value of all food consumed by slaves, calculated using standard estimates of slave consumption (given in Appendix B) and wholesale price data for major food items (given in Appendix C). The estimated value of food consumption per slave ranged from $17.51 to $18.63. The value of gross farm production is further reduced by $13.75 for each slave on the farm unit, an upper-bound estimate derived by Ransom and Sutch of the average value of clothing and housing consumed by plantation slaves. I use their upper-bound figures to minimize the income gap between slaveholding and slaveless operators. See Ransom and Sutch, *One Kind of Freedom*, pp. 210–2.
70. Cf. Atack and Bateman, *To Their Own Soil*, pp. 225–46.
71. Robert W. Fogel and Stanley L. Engerman, "The Economics of Slavery," in Robert W. Fogel, ed., *The Reinterpretation of American Economic History* (New York: Harper & Row, 1971), p. 338.

Tennessee farms was triple that in Middle Tennessee and very nearly quadruple that for East Tennessee. For purposes of comparison, it is revealing to note that the typical Tennessee slave annually consumed food and nonfood items (excluding housing) valued conservatively at $28 to $30.[72] Focusing strictly on the material rewards that farming afforded, we may say that free farm households in the plantation counties of West Tennessee enjoyed a standard of living five times higher than that of the average slave. In Middle and East Tennessee, the racial differential, although substantial, was far less dramatic. In those sections farm households earned incomes from basic crop and livestock production only one- to two-thirds greater in value than the average annual consumption of plantation slaves.

In addition to the vast differences in income levels that separated the regions, there were also significant patterns of variation within the farm populations of each, as Table 1.11 makes clear. Not surprisingly, within each section income per free household member was considerably higher in slaveholding households than in those without slaves – 75 to 80 percent higher in Middle and East Tennessee and approximately four times higher in West Tennessee, where slaveless farmers were limited largely to the bottom fourth of farm operators.

In every region there was also a major income differential that separated owner-operators from tenants. Owners' net incomes were nearly 2.5 times higher than those of tenants in East Tennessee, approximately three times higher in Middle Tennessee, and nearly four times higher in West Tennessee. Because the figures in Table 1.11 take into consideration the rent paid by tenants but not the rental income earned by owners, which cannot be determined from the census, the actual disparity between tenures was undoubtedly even greater. Keeping in mind the annual value of consumption among Tennessee bondsmen ($28 to $30 per capita, exclusive of housing), it is clear that tenants in both East and Middle Tennessee must have been forced to supplement their earnings from off-farm employment or else endure a material standard of living substantially beneath that of the plantation slave.[73]

The most significant line of demarcation, however, was neither slave-ownership nor tenure but degree of commercial orientation. In a seminal

72. Based on the estimated value of food consumed (see fn. 69) and the lower-bound estimates of Ransom and Sutch concerning slave consumption of clothing and medical services. See *One Kind of Freedom*, p. 211.
73. On the prevalence of off-farm employment, see Robert McGuire and Robert Higgs, "Cotton, Corn, and Risk in the Nineteenth Century: Another View," *Explorations in Economic History* 14 (1977):171–2.

essay published in the late 1960s, Morton Rothstein argued that it would be analytically useful to conceive of the Old South as a "dual economy." The region, he hypothesized, was divided between a "traditional" sector of slaves and nonslaveholding whites, who were but slightly concerned with production for the market, and a "modern" sector of slaveholders and prosperous slaveless yeomen, who were market-oriented in behavior and integrally connected to larger networks of exchange. Although acknowledging the pathbreaking nature of the essay, historians by and large have criticized the model as inadequate and the bipolarity that it posits as ahistorical. Nor, on the whole, is it applicable to the sample Tennessee counties studied here. For all three regions of the state, a close examination of farm production patterns reveals a smooth continuum from the subsistence-oriented to the highly commercial, not the stark dichotomy of traditional and modern households that Rothstein suggested.[74]

Unfortunately, although understandably rejecting the concept of a dual economy, historians have neglected Rothstein's insight that farmers isolated from the market economy – whether by choice or necessity – were by definition poor. Resurrecting Richard Hofstadter's "Myth of the Happy Yeoman," several recent studies of the nineteenth-century South have eloquently praised the "habits of mutuality" and "traditional social arrangements" that unified communities untouched (that is, untainted) by the market economy.[75] It would be foolish to argue that there were no social, cultural, and even psychological costs concomitant with market integration, but it is important also not to lose sight of the undeniable economic costs of persistent market isolation.

In Tennessee, households oriented toward market production – defined conservatively as those that sold at least half of all they produced – earned incomes far above those enjoyed by subsistence-oriented farmers (i.e.,

74. Morton Rothstein, "The Antebellum South as a Dual Economy: A Tentative Hypothesis," *Agricultural History* 41 (1967):373–82. See also North, *The Economic Growth of the United States*, pp. 101–55. In fairness, Rothstein did argue that the dual economy concept was likely most applicable to the cotton states of the Lower South. For criticism of the concept, see Bode and Ginter, *Farm Tenancy and the Census*, p. 5; Ford, *The Origins of Southern Radicalism*, pp. 57ff.; Schlotterbeck, "Plantation and Farm," p. 334.

75. For examples see Hahn, *The Roots of Southern Populism*, p. 29; Schlotterbeck, "The 'Social Economy' of an Upper South Community," p. 22; Weiman, "Petty Commodity Production in the Cotton South," pp. 80–95; and several of the essays contained in Hahn and Jonathan Prude, eds., *The Countryside in the Age of Capitalist Transformation: Essays in the Social History of Rural America* (Chapel Hill: University of North Carolina Press, 1985).

those who sold less than half).[76] According to these definitions, only about one-third of farmers in East and Middle Tennessee were market-oriented; significantly, they earned per-capita incomes from farm operations more than three times greater than those of their subsistence-oriented neighbors. In West Tennessee the disparity between groups was even greater. In contrast to the other two regions, farmers heavily involved in the market comprised over four-fifths of farm operators. Their incomes dwarfed by a factor of almost ten those earned by the remaining one-fifth who produced marginally for the market.

The most striking revelation is how little removed subsistence-oriented white farmers were from the standard of living of southern slaves. Indeed, only in Middle Tennessee did subsistence-oriented farmers earn incomes above the average value of slave consumption, and there the margin of difference was slight. The point is not that the subsistence-oriented were worse off materially than slaves, for unlike slaves, they could take advantage of off-farm opportunities to augment their incomes. What the income estimates do make clear is their absolute dependence on such additional income; without it they routinely earned less than the amount masters allocated to their chattel.

In light of these findings it is all the more difficult to accept the argument that, when free to choose, southern yeomen sought consciously to limit their involvement in larger networks of exchange. Rather, it seems probable that farmers welcomed the advantages that access to external markets offered (substantially higher incomes being one of these), while simultaneously striving to preserve traditional social arrangements. Furthermore, there were no compelling reasons why they should have seen the two goals as incompatible. In Tennessee, at least, farmers who committed heavily to the market were more likely to have been self-sufficient than their subsistence-oriented neighbors. Although they certainly feared the abuses of the market, it is likely that few farmers resisted the market *per se*. For Tennessee farmers who were unwilling or unable to participate substantially in the market economy, self-sufficiency was, to quote Rothstein, but a "delightful euphemism for rural poverty."[77]

76. Thus subsistence orientation is inferred from production patterns. It is worth noting that such orientation was not necessarily a product of conscious choice or preference.

77. Rothstein, "The Antebellum South as a Dual Economy," p. 375. This generalization held true for "self-sufficient" communities as well as for individual households. An evaluation of aggregate county-level data for the entire state indicates that mean income per capita among farm-operating households was $49.42 in coun-

Conclusion

This chapter began with a discussion of the heterogeneity of southern agriculture on the eve of the Civil War. The sample Tennessee counties varied greatly from region to region in several important respects, especially in dependence on slavery, a factor that contributed in turn to significant variations in work routine, overall scale of operations, structure of the farm population, emphasis on production for the market, and level of farm income.

At the same time, this analysis has revealed important similarities among the three regions of Tennessee. Despite the much greater prevalence of large landholdings in West Tennessee, the size of the typical farm operation did not vary dramatically across the state. In every section farmers generally produced enough food to provide for their own households and livestock. Finally, all across the state farmers typically marketed substantial proportions of their output without forsaking self-sufficiency. Although those either unwilling or unable to participate in larger networks of exchange were invariably poor, farmers with both the inclination and the opportunity to do so increased their income and in turn their material standard of living. The result was not decreased independence but increased wealth.

ties in which less than half of total agricultural production was marketed. The income level for the more commercially active counties was more than twice as high, $103.61. These are mean values, however, thus not strictly comparable to the median estimates presented for the sample counties. The published county-level data do not permit an estimation of median income levels, which were undoubtedly much lower for both categories.

2. "Honest Industry and Good Recompense": Wealth Distribution and Economic Mobility on the Eve of the Civil War

The preceding examination of agricultural patterns during the late ante-bellum period demonstrates the dangers of facile assumptions regarding southern heterogeneity. Although Tennessee's "grand divisions" admittedly differed in several striking respects – most notably in reliance on slavery, prevalence of large plantations, and dependence on cotton – any generalization stressing such differences would be apt to mislead in not one but two ways. It would likely exaggerate *inter*sectional diversity – minimizing similarities across the state in typical scale of operations and frequency of self-sufficiency, for instance – while glossing over important *intra*sectional differences – for example, the gross income disparities that separated market and subsistence-oriented farmers within all three sections.

Patterns of farm operations, however, are not the only yardstick of diversity among farm populations. Scholars who compare Black Belt and Upcountry areas frequently maintain that differences in the extent of plantation slavery also contributed to fundamental dissimilarities in socioeconomic structure.[1] Their argument rests on two reasonable but largely unproven propositions with regard to the South as a whole: first, that

1. For examples see Steven Hahn, *The Roots of Southern Populism: Yeoman Farmers and the Transformation of the Georgia Upcountry, 1850–1890* (New York: Oxford University Press, 1983), pp. 20ff; David Weiman, "Petty Commodity Production in the Cotton South: Upcountry Farmers in the Georgia Cotton Economy, 1840–1880" (Ph.D. diss., Stanford University, 1983), pp. 274–91; Ronald Eller, *Miners, Millhands, and Mountaineers: Industrialization of the Appalachian South* (Knoxville: University of Tennessee Press, 1982), pp. 3–38; and John Mitchell Allman III, "Yeoman Regions in the Antebellum Deep South: Settlement and Economy in Northern Alabama, 1815–1860" (Ph.D. diss., University of Maryland, 1979), pp. 2–3, 392.

slavery promoted a greater concentration of wealth than would have obtained otherwise, and second, that it restricted opportunities among nonslaveholding whites for economic advancement. By extension, local areas predominantly characterized by small farms and white labor should have exhibited more egalitarian distributions of wealth and higher levels of economic opportunity than did plantation districts.

Unfortunately, works that compare the social and economic structure of different areas within the South are virtually nonexistent. Most broadbased analyses have either treated the South as a whole or have been confined to a specific subregion such as the "Cotton South." Although valuable, these analyses have little to say about internal variation.[2] Potentially more helpful are numerous recent case studies of individual counties. Studied collectively they are suggestive, but inconsistency in research design and technique make comparison problematic.[3]

2. For example, Lee Soltow, *Men and Wealth in the United States, 1850–1870* (New Haven: Yale University Press, 1975), pp. 124–46; Gavin Wright, " 'Economic Democracy' and the Concentration of Agricultural Wealth in the Cotton South, 1850–1860," *Agricultural History* 44 (1970):63–93; Donald Schaefer, "Yeomen Farmers and Economic Democracy: A Study of Wealth and Economic Mobility in the Western Tobacco Region, 1850–1860," *Explorations in Economic History* 15 (1978): 421–37. Wright examines differences in wealth distribution by soil region, but confines his analysis to the "Cotton South." Two exceptions to the foregoing generalization are Albert W. Niemi Jr., "Inequality in the Distribution of Slave Wealth: The Cotton South and Other Agricultural Regions," *Journal of Economic History* 37 (1977):747–54; and Randolph B. Campbell and Richard G. Lowe, *Wealth and Power in Antebellum Texas* (College Station: Texas A&M University Press, 1977). Unfortunately, by focusing on the leading counties in production of various commodities, Niemi systematically excludes from consideration all but the most commercially developed sections of the South. Campbell and Lowe explore wealthholding variations in four distinct regions within the eastern two-thirds of Texas; despite differences in climate, topography, and soil type, the regions were relatively homogeneous in dependence on slavery and commitment to the cotton economy.

3. Among many, see Frank Jackson Huffman Jr., "Old South, New South: Continuity and Change in a Georgia County, 1850–1880" (Ph.D. diss., Yale University, 1974); John Thomas Schlotterbeck, "Plantation and Farm: Social and Economic Change in Orange and Greene Counties, Virginia, 1716 to 1860" (Ph.D. diss., The Johns Hopkins University, 1980); Carl H. Moneyhon, "Economic Democracy in Antebellum Arkansas, Phillips County, Arkansas, 1850–1860," *Arkansas Historical Quarterly* 40 (1981):154–72; William L. Barney, "Towards the Civil War: The Dynamics of Change in a Black Belt County," in Orville Vernon Burton and Robert C. McMath (eds.), *Class, Conflict and Consensus: Antebellum Southern Community Studies* (Westport, CT: Greenwood Press, 1982), pp. 146–72; Randolph B. Campbell, *A Southern Community in Crisis: Harrison County, Texas, 1850–1880* (Austin: Texas State Historical Association, 1983); Hahn, *The Roots of Southern Populism*; and J. William Harris, *Plain Folk and Gentry in a Slave Society: White*

This chapter presents a more systematic, explicitly comparative analysis through an investigation of *local* patterns of wealthholding and economic mobility in the eight sample Tennessee counties. Admittedly, the contours of property ownership do not by themselves delineate the social structure of a community; the distribution of wealth never corresponds exactly with the distribution of power, status, or influence. Nor does the extent of economic mobility – the acquisition or loss of wealth – serve as a perfect barometer of the openness or fluidity of a society.[4] Even so, as one prominent scholar contends, "in the antebellum era wealth appears to have been the surest sign of social, as well as of economic, position."[5] A clearer understanding of how wealth distribution and patterns of accumulation varied between plantation and small farm regions constitutes an essential step in the exploration of southern socioeconomic diversity.

Local Distribution of Wealth

"The tendency of things . . . in slave countries," the British economist J. E. Cairnes observed in 1863, "is to a very unequal distribution of wealth."[6] During the past two decades, numerous studies of wealthholding in the United States have upheld Cairnes's conclusion, although they have also shown that high concentrations of wealth were not peculiar to the antebellum South. On the contrary, a "very unequal distribution of wealth" appears to have been a typically American, rather than uniquely southern, characteristic.[7]

The fact that the distribution of wealth in the antebellum South *as a whole* was highly concentrated does not, of course, obviate the possibility that the distribution of wealth in certain types of localities may have been

Liberty and Black Slavery in Augusta's Hinterlands (Middletown, CT: Wesleyan University Press, 1985).

4. For a discussion of these points see Edward Pessen, "Social Mobility in American History: Some Brief Reflections," *Journal of Southern History* 45 (1979):165–84; and James A. Henretta, "The Study of Social Mobility: Ideological Assumptions and Conceptual Bias," *Labor History* 18 (1977):165–78.
5. Edward Pessen, "How Different from Each Other Were the Antebellum North and South?", *American Historical Review* 85 (1980):1119–49 (quotation on p. 1130).
6. John E. Cairnes, *The Slave Power: Its Character, Career & Probable Designs*, 2d ed. (London: Macmillan and Co., 1863; reprint ed., New York: Augustus M. Kelley, 1968), p. 76.
7. Peter H. Lindert and Jeffrey G. Williamson, "Three Centuries of American Inequality," *Research in Economic History* 1 (1976):69–123. See also references cited in fns. 2 and 3, in particular Soltow, *Men and Wealth in the United States*, and Wright, " 'Economic Democracy.' "

substantially less concentrated. This supposition is not borne out, however, by wealthholding patterns in Tennessee in 1860. Table 2.1 presents weighted averages by region of distributions at the county level of real estate, personal estate, and total wealth as recorded in the federal manuscript census. The figures in each section of the table are based on an analysis of every farm household in each of the sample counties.[8]

As Table 2.1 shows, the concentration of landownership was substantial in all three Tennessee regions. As popular stereotypes would suggest, it was greatest in the Black Belt counties of West Tennessee, where the top 5 percent of farm households owned nearly 46 percent of real wealth and the bottom half possessed only 3 percent of the total. More surprising is the discovery that levels of concentration were but slightly lower in Middle Tennessee, where plantation slavery was far less prevalent, and in East Tennessee, where it was virtually nonexistent. Although large landholdings were without question more prevalent in West Tennessee, and the concentration of real estate among farm *operators* greater as well, this was offset by the proportionally greater importance in Middle and East Tennessee of landless farmers and farm laborers not listed on the agricultural census. Indeed, the Gini coefficients of concentration vary but little among the three regions, and the coefficients computed for the latter two sections, though slightly lower than that for West Tennessee, are well within the range scholars have computed for other major plantation areas.[9]

8. Total N was 4,216 for East Tennessee, 5,551 for Middle Tennessee, and 2,059 for West Tennessee. Farm households, as defined in this study, include any household in which one or more members reported a farming occupation, whether planter, overseer, farmer, tenant, farm hand, or farm laborer. Defined and determined in this way, the average ratio of farm households to total households per county ranged from 0.77 in East Tennessee to 0.81 in West Tennessee.

   The distributions are constructed using estimates made by household heads to the census enumerator and are supposed to represent the "full market value" of property owned by that household, regardless of "any lien or encumbrance," and regardless of whether the property was actually situated within that enumeration district or county. See Carroll D. Wright and William C. Hunt, *The History and Growth of the United States Census*, S. Doc. 194, 56th Cong., 1st sess., 1900.

9. The Gini coefficient is a measurement of wealth concentration ranging from 0.0, indicating absolute equality in distribution, to 1.0, indicating complete concentration of an asset in the possession of one individual or household. Randolph B. Campbell calculated a coefficient of 0.68 for a single county in East Texas. Jonathan Wiener did not compute a Gini coefficient for his study of wealthholding in Marengo County, Alabama, but it is possible to derive from his published decile table a coefficient of 0.68 for that county also. See Campbell, *A Southern Community in Crisis*, p. 29; and Jonathan M. Wiener, *Social Origins of the New South: Alabama, 1860–1885* (Baton Rouge: Louisiana State University Press, 1978), p. 15.

Table 2.1 Distribution of wealth among free farm households, sample
Tennessee counties, 1860

	East	Middle	West
	Real Estate		
Percentage share of top 5%	39.1	40.4	45.5
Percentage share of bottom half	1.7	5.0	3.0
Gini coefficient	0.71	0.68	0.73
	Personal Estate		
Percentage share of top 5%	48.7	41.1	37.1
Percentage share of bottom half	6.7	5.6	3.0
Gini coefficient	0.70	0.69	0.71
	Total Estate		
Percentage share of top 5%	40.7	39.6	38.9
Percentage share of bottom half	4.9	6.4	3.8
Gini coefficient	0.69	0.67	0.70
Age-adjusted Gini coefficient	0.58	0.58	0.60

Source: Eight sample counties. See text.

The distributions of personal estate (movable wealth including but not limited to slaves) are also distinguished by their overall similarity. The ownership of personal property in the western Black Belt counties was not substantially more concentrated than in the mixed-farming section of Middle Tennessee or the subsistence-oriented counties of East Tennessee. Although the wealthiest households (top 5 percent) controlled a smaller proportion of the total in West Tennessee than in the other sections, this difference was offset by a greater concentration of wealth among the remainder of the farm population, as shown by the lower proportion of personal wealth owned by the bottom 50 percent.

As a final confirmation of intersectional uniformity, Table 2.1 reveals that the distribution of total wealth (the value of real and personal wealth combined) was not only highly skewed but also strikingly consistent across the state. In all three sections, the top 5 percent of wealthholders owned from 39 to 41 percent of local wealth, whereas the bottom half of local farm populations commanded only 4 to 6 percent of their communities' resources. Despite great variation across the state in organization of agricultural production, extent of commercial involvement, and degree of reliance upon slavery, farm communities in Tennessee from the Appalachian Mountains to the Mississippi River had nearly indistinguishable distributions of wealth.

Unfortunately, several wealth-related factors potentially obscure the social and economic significance of this impressive symmetry. To begin with there is the well-documented influence on cross-sectional wealth distribution of patterns of age structure and life-cycle accumulation, patterns that may have varied significantly across the state. Second, there were conspicuous differences among the regions of Tennessee in both the magnitude and composition of wealth (i.e., the importance of slave property), and these differences may have altered the social meaning of property ownership from region to region. Finally, it is generally true that cross-sectional comparisons may conceal important differences in the pace or extent of individual accumulation of wealth over time.

Wealth Distribution and the Life Cycle

Studies of a variety of times and places in the United States have consistently found a direct relationship between age of wealthholder and amount of wealth owned.[10] The same pattern obtained, with some variations, in all three Tennessee regions (see Figure 2.1). Although most pronounced in the western part of the state, in every section farmers typically accumulated wealth steadily throughout the course of the life cycle until old age, when their stock of wealth frequently declined, due in all likelihood to retirement or to the division of property among descendants prior to death.

Such clear trends of life-cycle accumulation complicate attempts to characterize and compare cross sections of local wealth distribution. Specifically, for each part of the state some unknown proportion of the observed inequality in 1860 was attributable to differences in mean wealth *between* age

10. For the eighteenth and nineteenth centuries, see Jeremy Atack and Fred Bateman, "Egalitarianism, Inequality, and Age: The Rural North in 1860," *Journal of Economic History* 41 (1981):85–93; Campbell and Lowe, *Wealth and Power*, pp. 57–9; Harris, *Plain Folk and Gentry in a Slave Society*, p. 85; James A. Henretta, "Families and Farms: Mentalité in Pre-Industrial America," *William and Mary Quarterly* 35 (1978):5–8; Alice Hanson Jones, *Wealth of a Nation to Be: The American Colonies on the Eve of the Revolution* (New York: Columbia University Press, 1980), pp. 213–14, 381–8; Soltow, *Men and Wealth in the United States*, pp. 27–31, 69–74, 105–11; and Weiman, "Petty Commodity Production in the Cotton South," pp. 301–5.

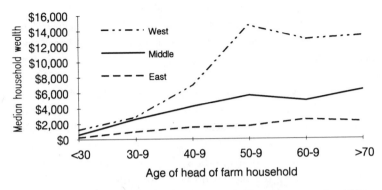

Figure 2.1   Wealth accumulation by age, sample Tennessee counties, 1860

groups – what might be labeled "age-wealth stratification" – whereas the remainder primarily reflected differences in individual wealth *within* age groups. If the latter category of inequality is arguably an "undemocratic" trait, the former is theoretically consistent with even the most "egalitarian" society.[11]

Table 2.1 presents "age-compensated" Gini coefficients of total wealth concentration that adjust statistically for the influence of life-cycle effects by employing a procedure developed first by Morton Paglin and modified slightly by Jeremy Atack and Fred Bateman.[12] The adjusted coefficients reveal that from one-third to two-fifths of the observed inequality in each

11. Robert Gallman makes this point forcefully in "Professor Pessen on the 'Egalitarian Myth,' " *Social Science History* 2 (1978):194–207.
12. Morton Paglin, "The Measurement and Trend of Inequality: A Basic Revision," *American Economic Review* 65 (1975):598–609; Jeremy Atack and Fred Bateman, "The 'Egalitarian Ideal' and the Distribution of Wealth in the Northern Agricultural Community: A Backward Look," *Review of Economics and Statistics* 63 (1981): 124–9. Paglin noted that the traditional Gini coefficient is computed with reference to a hypothetical standard of perfect equality in which every individual, regardless of age, controls an equivalent share of total wealth. Rejecting such an untenable standard, Paglin proposed instead that scholars employ as a new egalitarian ideal a society in which individuals would exhibit equivalent lifetime accumulation; at a given point in time individual wealth would indeed vary by age, but within age groups all individuals would command equal shares. In technical terms, the 45 degree line of perfect equality would be replaced by a new Lorenz-type, age-adjusted reference curve based on the empirically determined age–wealth profile of the population under consideration. Following Atack and Bateman, the age-compensated Gini is computed to equal the ratio of the area between the age-adjusted reference curve and the Lorenz curve to the area under the age-adjusted curve.

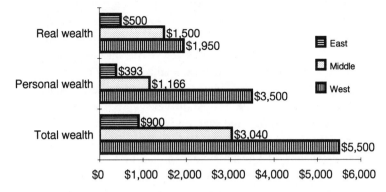

Figure 2.2 Median wealth of free farm households, sample Tennessee counties, 1860

region stemmed directly from the age–wealth relationship.[13] The unadjusted coefficients, in other words, substantially overstate the extent of inequality in rural Tennessee, a significant proportion of which was a direct function of the life cycle. Significantly, however, they do not exaggerate the degree of similarity among the three regions. Even after controlling for age–wealth stratification, the distribution of wealth continues to exhibit an extraordinary uniformity across the state.

Disparity in Levels of Wealth

Although interregional differences in life-cycle effects are apparently insignificant, the demonstrably large differences across the state in the average level of wealth are less easily dismissed (see Figure 2.2). Median total wealth per farm household in the Black Belt counties of West Tennessee was nearly double the figure for Middle Tennessee and more than six times that for the Appalachian counties in the eastern part of the state. As a direct result, the absolute difference between rich and poor was much greater in West Tennessee as well. For example, members of the local agricultural elite in East Tennessee (defined as the top 5 percent of farm households) typically owned several hundred acres of land and five to six slaves. At the other end of the state, in contrast, elite planters numbered

13. The assumption of perfect equality within age groups produces Gini coefficients – "age Ginis" according to Paglin's terminology – of 0.28 for East Tennessee, 0.22 for Middle Tennessee, and 0.25 for West Tennessee. Compare with the standard Lorenz Ginis presented in Table 2.1.

their acres in the thousands and owned on the average more than sixty slaves.[14] At the top of the economic pyramid for the western section was Haywood County's James Bond, who in 1860 owned five plantations totaling 17,000 acres. In 1859 Bond worked 220 slaves on those plantations to produce more than 1,000 bales of cotton and almost 22,000 bushels of corn (roughly one-fourth of the total corn produced in East Tennessee's Johnson County). Together with his wealthy neighbor Nathan Adams, who owned more than 7,000 acres and 200 slaves, these two West Tennessee slaveholders commanded in value of total wealth one-fifth as much as the entire farm population of Greene County in the East Tennessee Valley (2,372 households) and actually exceeded in wealth the combined holdings of all 681 farm households in the remote mountain county of Johnson.

It is reasonable to ask whether such sizable differences in absolute level of wealth in some way altered the social meaning of wealth distribution among the three regions, despite their structural similarity. Given such inordinate dissimilarity in wealth level, how significant could it be that the distribution of total wealth in East and West Tennessee was almost identical? To pose the question more broadly, in nineteenth-century rural America, which had greater social implications for local community and individual consciousness, the extent of *absolute* or of *relative* inequality?

Absolute Versus Relative Inequality

Studies of the Old South rarely address both absolute and relative inequality, choosing instead – more often subtly than overtly – to concentrate on one or the other. Analyses of particular subregions – the Cotton South, the Georgia Upcountry, or the Alabama Black Belt, for example – often pay at least some attention to relative inequality. (Certainly any study employing Gini coefficients has done so; by definition the statistic is a relative measurement.) In comparing subregions, however, scholars have more commonly concentrated, implicitly if not explicitly, on variations in absolute inequality. Their focus is revealed in a penchant for ill-defined terms such as *planter* or *yeoman*, labels that historians have struggled to define for years without success. The unstated assumption has been that, once categorical definitions for individuals are agreed on – presumably on the

---

14. For an extensive discussion of the Tennessee agricultural elite, see Robert Tracy McKenzie, "Civil War and Socioeconomic Change in the Upper South: The Survival of Local Agricultural Elites in Tennessee, 1850–1870," *Tennessee Historical Quarterly* 52 (1993):170–84.

basis of levels of land and slave ownership – the task of classifying and comparing communities or regions should be simple and straightforward. To give an example, if one defines a planter as an individual owning twenty or more slaves, then one out of every six farmers in West Tennessee was a planter, whereas in East Tennessee the proportion was less than one in 500. By definition the former region qualified as a major plantation area and the latter as a yeoman enclave.

Such comparisons highlight major variations in level of wealth between regions, and are appropriately used for that purpose, but by relying on absolute rather than relative definitions of economic status they inevitably obscure the degree of differentiation within the regions themselves. Evaluated in relative terms, economic distinctions were no less pronounced in Middle and East Tennessee than within the Black Belt counties to the west. It is important to stress this point particularly in regard to the easternmost section of the state, given the common perception of Southern Appalachia as akin to a Jeffersonian utopia. Despite pretentious nicknames such as "the Switzerland of America" or "the poor man's rich land," in reality the area exhibited much the same variety and range of economic circumstances as found in wealthier regions.[15]

In East Tennessee's Greene County, for example, just under two-fifths of farm household heads worked as tenants or laborers on land they did not own. Although a good proportion of these owned modest amounts of personal property, earned a decent income from their labors, and had respectable chances of acquiring land of their own in the future, others can be classified only as desperately poor. Among the latter were members of the Morgan clan from southwestern Greene County, six related families who, according to a local census official, "live[d] up on the side of the Alleghany mountains in a very Savage State." With obvious disdain, the enumerator explained that "these Morgans . . . raise a little corn by digging up the ground with a hoe[.] Some times Some of them come down in the County and work by the day and bring corn to sell." Only a few miles away, but a world apart in social and economic status, lived Dr. Alexander Williams, known to Greene Countians as the "Grand Duke of Lick Creek." Williams owned two fine farms in the western part of the county, forty-

---

15. For contemporary nicknames, see Eugene W. Hilgard [Special Agent], *Report on Cotton Production in the United States* (Washington, DC: Government Printing Office, 1884), p. 410; Hermann Bokum, *The Tennessee Handbook and Immigrant's Guide* (Philadelphia: J. B. Lippincott & Co., 1868), p. 8. For an example of the common view of southern Appalachia as unusually egalitarian, see Eller, *Miners, Millhands, and Mountaineers*, Chap. 1.

four slaves, and, according to the local newspaper, "the most elegant and desirable residence to be found in East Tennessee," a three story mansion in the county seat of Greeneville.[16]

Clearly, categorizing Greene County or others like it as a yeoman stronghold conceals as much as it reveals of the socioeconomic reality. Neither Alexander Williams nor other members of the agricultural elite in East and Middle Tennessee could match the extremes of wealth exhibited among the largest slaveholding farmers of West Tennessee. None could rival James Bond's 17,000 acres, and few could adorn their wives with diamonds in the manner of a Nathan Adams.[17] Even so, evaluated within their own context, the wealthiest farmers in both sections towered above their local communities just as impressively. If wealthholding categories are to be employed to delineate the social divisions of local communities, those categories should be defined within the context of the local wealthholding structure, not based on absolute definitions intended to transcend geographic and temporal bounds.

Such an approach should lead to a new understanding of how local socioeconomic structure varied across the South. It might also challenge common inferences about individual social consciousness. Underlying historians' emphasis on absolute inequality is the implicit assumption that rural Southerners formed their perceptions of economic position with reference to some absolute regionwide standard of comparison. From this premise it logically follows that, because the socioeconomic spectrum was considerably narrower in absolute terms outside the Black Belt, the extent of community stratification must have been less there as well.

Although the line of reasoning makes sense, the presupposition on which it is founded is arguable. Almost a half century ago Fabian Linden ques-

16. United States Census Office, Eighth Census of Population [1860], microfilm publication series #653, roll #1252, p. 405b; David Warren Bowen, *Andrew Johnson and the Negro* (Knoxville: University of Tennessee Press, 1989), p. 19; Greeneville *New Era*, 9 September 1865. During the Civil War the Williams mansion became famous as the headquarters of Confederate general John Hunt Morgan on the evening that he was surrounded and ultimately killed by Union soldiers. The home is described in detail in James A. Ramage, *Rebel Raider: The Life of General John Hunt Morgan* (Lexington: University Press of Kentucky), pp. 232–3.

17. Grace Adams's jewelry collection included "Three diamond finger Rings, three Setts of diamond Ear Rings, Four diamond Breast Pins, One diamond Locket and Necklace, One large long gold chain for the Neck, Four gold Bracelets, Two Watches and Chains, [and] One Sett of Cameo Ear Rings and Pin." See will of Grace Arrington Howell Adams, 3 December 1877, in Tennessee State Library and Archives, Haywood County Records, Office of the County Court Clerk, Wills, mf. roll #47, vol. E, pp. 363–4.

tioned it, arguing that "the status, prestige, and influence of the farmer depended not so much on the absolute size of his acreage, but more on his rank within his immediate universe, on his place in the 'pecking order' within the circumference of his own barnyard."[18] If correct, his point is crucial, for as Mills Thornton has persuasively argued, "the definition of southern reality . . . must begin with the southern perception of reality."[19] In actuality, whether local pecking orders or interregional disparities were more important in shaping popular perceptions likely depended on the issue at hand. If the question involved intersectional struggles for power in state politics, then the vast disparity in absolute wealth among the regions was likely the prevailing factor. On the other hand, one need not exaggerate the rural isolation of the nineteenth-century South to recognize that subtle distinctions shaped by relative inequality probably informed, if they did not determine, individual perceptions of local stratification. The point is not to argue that individual perceptions of local social structure were identical across the state but, rather, to question whether the differences that may have existed bore any necessary connection to variation in absolute levels of overall wealth.

Disparity in the Composition of Wealth

In addition to disparities in wealth level, substantial differences among the Tennessee regions in the composition of wealth – specifically the proportional importance of slave property – cloud the significance of their similar distributions of wealth. Tennessee Civil War veterans interviewed in their old age differed concerning the distinctiveness of slave property as opposed to other forms of wealth. Several professed to see little difference. Charles Ross, the son of a slaveless landowner in West Tennessee, recalled that slaveholders in his community "all ways had the advantage just as the money man has it today." Middle Tennessean W. A. Vardell, whose family owned neither land nor slaves, described relations between slaveholders and

---

18. Fabian Linden, "Economic Democracy in the Slave South: An Appraisal of Some Recent Views," *Journal of Negro History* 31 (1946):140–89 (quotation on p. 164). Although scholars frequently cite this article as a devastating critique of the Owsley School, they have largely overlooked Linden's implicit suggestion that patterns of wealthholding varied slightly across the South. In the same paragraph containing this quotation, Linden alluded to "the striking similarity of the socioeconomic pattern of ownership in all the regions studied.".

19. J. Mills Thornton III, *Politics and Power in a Slave Society: Alabama, 1800–1860* (Baton Rouge: Louisiana State University Press, 1977), p. 479.

nonslaveholders as "about like money man and the non mony man now, negro then, mony now." A veteran whose father had owned thirty slaves in a nearby county echoed this sentiment, stating simply that "there was no difference except as exists now between the wealthy and the poor." Edmund Gannaway, son of a small slaveholder in the East Tennessee Valley, elaborated at greater length, denying that "welth in slaves gave men the bighead more than welth in land or monney or any thing[.]" As Gannaway explained, "some people cant stand riches. [A]n automobile will spoil a whole family sometimes."[20] Such statements recognized the reality that slaveownership and wealth went hand in hand in the Old South. Most families of above average wealth owned slaves, and all families that owned slaves were by definition of above-average wealth.[21]

A minority of Tennessee veterans, however, voiced sentiments to suggest that the presence of slaves in their communities peculiarly influenced their perceptions of local social structure and the prevalence of white economic opportunity therein. "A renter had no chance to save anything," West Tennessean William Beard complained. Unaware that slaveownership was as much a reflection as a cause of wealth, he was convinced that "slave holders were the only men that could save enough money to do anything." Anderson Roach of Grainger County in East Tennessee attributed his family's poverty to the influence of local slaveholders who "kept the poor class of people down as much as possible." A. J. Ferrell, who grew up in Middle Tennessee in a tenant's one-room log cabin, denied that slaveholders in his community actively discouraged the ambitious poor man but admitted that "if they had not owned slaves a man working as I was could have secured better wages." John Welch also blamed low wages for his family's destitution – in 1860 they owned house furniture and two cows, worth in all about $60 – but noted as well that in his section of western Tennessee "the slave holders owned nearly all the land and they wanted to keep it for their children."[22]

Such responses were not typical among Tennesseans interviewed. The

---

20. Gustavus W. Dyer and John Trotwood Moore, comps., *The Tennessee Civil War Veterans Questionnaires* (Easley, SC: Southern Historical Press, 1985), vol. 5, pp. 1881–2, 2095–6, 2179–80; vol. 3, p. 878.
21. In the Cotton South the average slaveholder was ten times wealthier than the average nonslaveholder. See Gavin Wright, *The Political Economy of the Cotton South: Households, Markets, and Wealth in the Nineteenth Century* (New York: W. W. Norton & Company, 1978), pp. 35–6.
22. Dyer and Moore, comps., *Tennessee Veterans Questionnaires*, vol. 1, pp. 299–300, 106–7; vol. 2, pp. 806–7; vol. 5, pp. 2156–7.

vast majority maintained that economic opportunities had been good in their antebellum communities, and some even made extravagant claims of extensive mobility.[23] James Williams recalled that in Middle Tennessee "the opportunities was good for a poor man and a lot of poor men got rich." East Tennessean Elisha Taylor claimed to "have known men who worked at 25 cents per day on farm that has left to their heirs land worth fifty thousand." Reflecting on conditions in West Tennessee, William Tucker noted "that the rich boy in many cases would loose and become poor, while the poor boy had become rich."[24] Still, the number of respondents who remembered chances for upward mobility as poor and who attributed that condition to the influence of slavery is too large to be ignored.

Patterns of Mobility over Time

In comparing local wealthholding structures across Tennessee, then, a final variable that should be considered is the extent of economic mobility within those structures. "The welthy was all the time seeking more," veteran William Orr recalled, "while the other fellow was trying to clime up thare two."[25] Orr's insight suggests that the typical Tennessee farmer must have evaluated the local pecking order by weighing both current position and future prospects. Perceptions were formed not only with reference to present patterns of property ownership, in other words, but also with an eye to potential upward mobility. Certainly, contemporary critics of southern society regularly linked these factors. When the British economist J. E. Cairnes wrote of the "very unequal distribution of wealth" within the South, or when Hinton Rowan Helper, a Southerner himself, complained that "a small minority conceited and tyrannical" dominated the region's economy, both men assumed that the great concentration of wealth in the South relegated the vast majority of whites to a permanently inferior status at the bottom of the economic ladder.[26]

Northern politicians shared these views. Two pervasive themes in the Republican Party's critique of the South were the characterization of the

23. Fred Arthur Bailey, *Class and Tennessee's Confederate Generation* (Chapel Hill: University of North Carolina Press, 1987), p. 158, Table 19.
24. Dyer and Moore, comps., *Tennessee Veterans Questionnaires*, vol. 5, pp. 2199, 2030–1, 2076–7.
25. Ibid., vol. 4, p. 1666.
26. Cairnes, *The Slave Power*, pp. 76, 95–7 and passim; Hinton Rowan Helper, *The Impending Crisis of the South: How to Meet It* (New York: A.B. Burdick, 1860), p. 25.

region as dominated by a tiny oligarchy and the conviction that social mobility was nonexistent there. Significantly, when Republicans held up the North as a model for southern emulation, it was not an allegedly egalitarian distribution of wealth that they emphasized but, rather, a purported equality of opportunity for economic advancement. "Improvement of condition ... is the order of things in a society of equals," Abraham Lincoln assured northern audiences. The essence of northern superiority rested on one fundamental principle: "The hired laborer of yesterday labors on his own account today, and will hire others to labor for him tomorrow."[27]

Despite its importance, historians still know very little about economic mobility in the Old South and almost nothing about how it varied between plantation and small-farm regions.[28] Several factors may help to account for this surprising dearth of knowledge. To begin with, the longitudinal analysis necessary to determine patterns of individual mobility over time is a tedious and time-consuming process. Given the limited resources available to most researchers, it is not surprising that studies commonly emphasize the relatively more accessible data concerning the static distribution of wealth.

Second, historians may have been influenced by current econometric literature, much of which appears to rest on the unspoken premise that cross-sectional wealth data in isolation constitute a true measurement of "economic democracy." Economists have debated extensively whether the Gini coefficient or some other statistic is the better measurement of inequality. Although they differ concerning the technical construction of an appropriate index, in discussing the "normative significance" or "welfare implications" of their debate they reveal a common foundational presupposition: After all complicating variables are accounted for (age, household

27. On the Republican critique of the South see Eric Foner, *Free Soil, Free Labor, Free Men: The Ideology of the Republican Party before the Civil War* (New York: Oxford University Press, 1970), pp. 40–72. Lincoln is quoted in Richard Hofstadter, *The American Political Tradition and the Men Who Made It* (New York: Alfred A. Knopf, 1948), pp. 133–4.

28. As late as 1982, a historiographical survey of mobility studies cited eighteen works employing record linkage techniques; of these, only one concerned a southern community, and that study dealt with Atlanta, hardly representative of the rural South. See Donald H. Parkerson, "How Mobile Were Nineteenth-Century Americans?", *Historical Methods* 15 (1982):107. This disparity is slowly beginning to change. Studies pertaining to the South that do explore patterns of economic mobility include Donald Schaefer, "Yeomen Farmers and Economic Democracy"; and Donald L. Winters, "The Agricultural Ladder in Southern Agriculture: Tennessee, 1850–1870," *Agricultural History* 61 (1987):36–52.

size, schooling, gender, etc.), there is some *single* level of wealth inequality that is most consistent with social justice and best reflects the social ideal. The assumption is dubious in the extreme, however. There is nothing inherently democratic or just about a given pattern of wealth ownership, and cross-sectional wealth data in isolation are never adequate indicators of social stratification.[29]

Finally, some scholars have probably eschewed mobility analyses for theoretical or ideological reasons. As mentioned in Chapter 1, a number of historians now reject what might be called the "entrepreneurial school" of agricultural history, which posits a near universal commitment among American farmers to profit maximization and wealth accumulation. Decrying the implicit "primacy of individualist values" on which mobility studies are often predicated, they question whether tests for mobility are really appropriate for population groups in which year-to-year subsistence and the financial security of the family were supposedly overriding goals. Regarding the South specifically, they reject the portrayal of white Southerners as liberal capitalists and describe instead a society dominated by "country republican" values – commitment to family, community, and economic independence – all objectives that mobility studies are ill-equipped to measure.[30]

There is much common sense (if also a measure of wishful thinking) in this challenge to the entrepreneurial interpretation of rural society. Although proponents may exaggerate the anticapitalist ethos of communities not yet "invaded and conquered by the logic and relations of commodity

29. For examples of the technical literature, see especially the responses of numerous authors to Paglin's 1975 article in "The Measurement and Trend of Inequality: Comment," *American Economic Review* 67 (1977):497–519. Works that rightly question this foundational premise include Gallman, "Professor Pessen on the 'Egalitarian Myth' "; and J. R. Kearl and Clayne L. Pope, "Wealth Mobility: The Missing Element," *Journal of Interdisciplinary History* 13 (1983):461–88.

30. James A. Henretta has written two seminal essays in this regard, "The Study of Social Mobility," quotation from p. 165; and "Families and Farms." Regarding the South specifically, see Barbara Jeanne Fields, "The Nineteenth-Century American South: History and Theory," *Plantation Society* 2 (1983):7–27; Hahn, *The Roots of Southern Populism*, pp. 29–39; Forrest McDonald and Grady McWhiney, "The South from Self-Sufficiency to Peonage: An Interpretation," *American Historical Review* 85 (1980):1095–118; John Thomas Schlotterbeck, "Plantation and Farm," p. 22; Harry L. Watson, "Conflict and Collaboration: Yeomen, Slaveholders, and Politics in the Antebellum South," *Social History* 10 (1985):273–98; and David F. Weiman, "Petty Commodity Production in the Cotton South," pp. 8ff. On country republicanism, see Lacy K. Ford Jr., *Origins of Southern Radicalism: The South Carolina Upcountry, 1800–1860* (New York: Oxford University Press, 1988), pp. 49–51.

production," they are on target in observing that individualist values were never universal in the nineteenth-century South.[31] This caveat is especially relevant when assessing attitudes with regard to wealth in slaves. If in West Tennessee "every non slave holder was working to become a slaveholder," as a veteran from that section believed, the goal was undoubtedly less common outside of the Black Belt. In East Tennessee, for example, limitations of climate, soil, and access to market prevented many farmers from the heavy commitment to cash-crop agriculture that made slaveownership profitable. As veteran John Howell recalled, "there were quite a number of people [in East Tennessee] who did not own slaves because of the fact it was very questionable about it paying to own them." Even when farmers could afford to purchase a slave, in other words, it did not always make economic sense to do so. Nor is it certain that all Tennessee farmers even wanted to own slaves. East Tennessean Anderson Landers, for example, was remembered by his son as "immensely Southern" but opposed to slavery.[32]

Rural white Tennesseans may have differed in their desire to own slaves, but their hunger for land was little short of universal. Admittedly, the social meaning of such aspirations may have varied; the successful acquisition of a farm may have been either the means to further objectives or an end in itself – the first step in a lifelong quest for riches or the final triumph in a struggle for independence. Either way, however – means or end – landless farmers yearned to own land.

The extent to which farmers who already owned land would have longed to acquire more is less evident. Because of a firm commitment to country republican values, many landowners may have been indifferent at best, at least if their farms were productive enough to sustain family subsistence from year to year, profitable enough to provide for their old age, and large enough to subdivide among their children. Mobility studies offer little insight when applied to communities characterized by such truly independent smallholders.

31. Steven Hahn, "The 'Unmaking' of the Southern Yeomanry: The Transformation of the Georgia Upcountry, 1860–1890," in Hahn and Jonathan Prude, eds., *The Countryside in the Age of Capitalist Transformation: Essays in the Social History of Rural America* (Chapel Hill: The University of North Carolina Press, 1985), p. 180. For a sharply critical overview of historians' recent emphasis on republicanism in the nineteenth century, see John Patrick Diggins, "Comrades and Citizens: New Mythologies in American Historiography," *American Historical Review* 90 (1985): 614–38.

32. Dyer and Moore, comps., *Tennessee Veterans Questionnaires*, vol. 5, p. 2150; vol. 3, p. 1166; vol. 4, p. 1323.

It is doubtful, however, whether many such communities ever existed for more than a generation in the nineteenth-century South. This was not due necessarily to the absence of an appropriate precapitalist mentality, which may have been common enough. Rather, if the sample counties were at all representative, the sticking point was the extraordinary rarity of the structural characteristics (i.e., patterns of landholding and production) necessary to sustain the single-minded pursuit of precapitalist objectives. Simply put, by 1860 large numbers of Tennessee farmers in all regions of the state could not have provided for the annual subsistence of their households and the long-term independence of their lineage without additional land.

To begin with the most obvious, between 25 and 40 percent of all heads of farm households owned no land whatsoever and, as tenants or farm laborers, were entirely dependent on others for their livelihood (refer to Table 1.3). A far smaller proportion, roughly 3 to 5 percent of the total, were tenants in part, that is, owner-operators who owned only a portion of their farms and rented the remainder, apparently needing or desiring more acreage than they could afford to purchase.[33] Other owner-operators, estimated conservatively at 7 to 12 percent of agricultural households, owned farms too small to sustain even minimal standards of living.[34] Finally, many of the remainder – landowners whose farms provided a fairly comfortable subsistence – owned plots far too small for feasible subdivision among their offspring. Even if only farms of fewer than fifty improved acres are deemed too small for subdivision, it would still appear that, at the very least, one-half to two-thirds of Tennessee farmers lacked the minimum resources necessary for individual and familial independence.[35]

33. The exact percentages were 3.4 in East Tennessee, 5.2 in Middle Tennessee, and 5.4 in West Tennessee. For a discussion of tenants in part, see Chap. 1, fn. 26.

34. I define a minimum standard of living to equal the upper-bound estimates of Roger Ransom and Richard Sutch concerning per capita consumption among slaves of food, clothing, and medical supplies (adjusted for Tennessee prices). Because small-holders likely supplemented their major crop and livestock outputs with fish, wild game, and garden and orchard products (sources of income not reflected in the census manuscripts), I have defined as deficient only those farm-operating households with per capita incomes less than 75 percent of the Ransom and Sutch estimates of slave consumption. See Ransom and Sutch, *One Kind of Freedom: The Economic Consequences of Emancipation* (New York and London, 1977), pp. 210–12. The percentage of all farm households in this category was 11.8 in East Tennessee, 11.5 in Middle Tennessee, and 7.0 in West Tennessee.

35. The proportion of total agricultural households who owned farms of at least fifty improved acres that met minimal subsistence standards (see fn. 34) was only 34.5 percent in East Tennessee, 40.3 percent in Middle Tennessee, and 52.0 percent in West Tennessee.

In sum, a stark dichotomy contrasting capitalist and precapitalist mentalities is ill-suited to the antebellum countryside. Undoubtedly, many white Southerners dedicated their lives preeminently to the quest for riches – "trying to clime up thare two," to borrow William Orr's phrase – while others devoted themselves single-mindedly to noneconomic, country republican goals. Others still, however, perhaps even the majority, worked hard to accumulate wealth *while* maintaining a strong devotion to family and community, objectives that were more complementary than competing. Recognition of the complexity of human motivation does not render evidence concerning wealth accumulation irrelevant, however. (If anything, it becomes all the more intriguing.) Rather, the historian is enjoined to approach the evidence with circumspection and a sensitivity to the complicated and sometimes ambiguous meaning of property accumulation in nineteenth-century rural America. Whether one posits a region of liberal capitalism, country republicanism, or something in between, evidence regarding economic mobility, when interpreted cautiously, offers important insight into the socioeconomic bases of the Old South and constitutes an indispensable supplement to cross-sectional evidence of wealth distribution.

Geographic Mobility

A major obstacle to almost any study of economic mobility in the nineteenth century is the constant and extensive turnover of population that characterized all regions of the country. As a consequence, studies necessarily focus on individuals who remained, or "persisted," in the same location over an extended period. Regrettably, it is typically impossible to learn why men and women left a particular locality, how they fared after reaching their destination, or what their departure signified about the nature of economic opportunity in the communities they left behind. High rates of migration especially complicate comparative studies of economic mobility, particularly when the rates of migration varied largely among the populations of interest. In Tennessee rates of geographic persistence differed substantially from region to region (see Table 2.2).[36] At the two extremes, just over one-third of farm household heads in West Tennessee

36. The following discussion and the figures presented in Tables 2.2 to 2.6 are based on a persistence analysis of a random sample of 3,901 heads of farm households from the eight sample counties. The household heads were selected from the 1850 census according to a random, stratified procedure and traced to the census of 1860. For a discussion of the sample and matching technique, see Appendix B.

Table 2.2 Geographic persistence among heads of free farm households, sample Tennessee counties, 1850-1860 (percentages)

	East	Middle	West
All farm households	55.8	47.8	36.6
Farmowners	61.5	51.5	44.7
Farm tenants	53.7	41.6	28.5
Farmers without farms	44.4	34.1	17.7

Source: Eight sample counties. See text.

remained in the same county throughout the 1850s, whereas well over one-half in East Tennessee did so.[37]

In the absence of extensive qualitative sources that document individual motives, any conclusion regarding the implications of migration would be overly speculative. Undoubtedly, many farmers moved for economic reasons.[38] The best evidence for this is indirect: In each region there was a strong correlation between landownership and geographic persistence, as Table 2.2 shows. When farmers owned the land they worked, they were much less likely to leave, a distinction that was greatest in West Tennessee, where landowners were twice as likely to persist as were the landless. All other things equal, the much lower persistence rate among landless farmers in the Black Belt suggests that they viewed their chances for land accumulation far more negatively than did farmers elsewhere in the state who suffered less from the competition of slave labor. In reality, of course, all other things were not equal, and the decision by a Tennessee farmer to leave his home was undoubtedly shaped by a complex array of factors that included – but was not limited to – economic considerations.[39]

37. Compare with the more uniform rates historians have found for other rural counties in the Deep South during the 1850s: for Clarke County, Georgia – 41.2 percent (based on all adult males); Dallas County, Alabama – 38.7 percent (based on all household heads); and Harrison County, Texas – 35.6 percent (based on all household heads). See Huffman, "Old South, New South," p. 35; Barney, "Towards the Civil War," pp. 146–72; and Campbell, *A Southern Community in Crisis*, p. 382.

38. Among Tennessee veterans interviewed in their old age, approximately two-thirds of those who had moved after the Civil War claimed to have done so in search of economic opportunity. See Bailey, *Class and Tennessee's Confederate Generation*, p. 163.

39. Peter Uhlenberg, "Non-Economic Determinants of Non-Migration: Sociological Considerations for Migration Theory," *Rural Sociology* 38 (1973):296–311.

It would be misleading, then, to attribute all decisions to migrate to pessimistic appraisals of the extent of opportunity afforded by the local social structure. Some farmers must have pulled up stakes simply because of sheer restlessness, others because they were discouraged by crop failures, still others in order to exchange wooded hills for level prairies.[40] A West Tennessee slaveholder confessed as much to his diary after spending most of the winter of 1854 clearing trees from his farm and becoming "heartily sick and tired" of it. "I am completely broken down heaving and setting at logs," he complained. "I must see Texas – I want prairie land."[41]

It would also be wrong to assume that a farmer's decision to remain in a community necessarily reflected a positive estimation of economic chances. The per-capita income figures given in Chapter 1 press home this point (see Table 1.11). For example, if 1859 figures are at all indicative, tenant farmers in East Tennessee earned incomes during the 1850s less than half as large as were earned by tenants at the other end of the state but were almost twice as likely to stay put. Clearly, factors other than profit maximization figured in their decision to stay or leave.

The high rates of persistence among East Tennessee farmers may have been a reflection less of the extent of economic opportunity in the region than of the degree to which mountain farmers weighed noneconomic factors in their decisions to remain. Numerous scholars have hypothesized that kinship networks were more important in Upcountry areas than in the more commercially developed Black Belt.[42] If so, then the pull of established family ties may have deterred many from pursuing their economic self-interest elsewhere.

Aversion to the risk associated with migration also may have discouraged many farmers from leaving, for picking up and moving one's family was an expensive undertaking. In an ironic way, the far greater incomes earned by tenants in West Tennessee may have facilitated migration (to less settled areas where available land was more abundant), whereas the far lower incomes among the landless elsewhere hindered migration and possibly bound numerous households to communities they might otherwise have

40. Merle Curti, *The Making of an American Community: A Case Study of Democracy in a Frontier County* (Stanford, CA: Stanford University Press, 1959), pp. 65–6.
41. Harrod C. Anderson Diary, Louisiana State University Library, entry for 21 February 1854.
42. See Eller, *Miners, Millhands, and Mountaineers*, pp. 28–32; Hahn, *The Roots of Southern Populism*, pp. 22–39; Henretta, "Families and Farms," pp. 19–25; Schlotterbeck, "Plantation and Farm," p. 22; and Weiman, "Petty Commodity Production in the Cotton South," p. 8.

abandoned. At any rate, it is worth noting that, in the larger American context, it is the extremely high persistence of farmers in eastern Tennessee rather than the extremely low persistence of western farmers that is anomalous.[43]

Economic Mobility

Migrating farmers, then, disappeared by and large from the historical record and their fate is unknown. The following discussion centers on those who remained; sufficient information survives to allow a sketchy outline of their experiences. In assessing the economic fortunes of these "persisters," it is useful to distinguish between two types of economic mobility. The first, here designated absolute economic mobility, consisted of movement between either the ranks of slaveless and slaveowner or of landless and landowner. As members of a society in which "social divisions . . . were essentially wealthholding categories," farmers who crossed these boundaries underwent clear-cut changes in socioeconomic status.[44] The second type, relative economic mobility, involved farmers whose wealth level changed significantly but who did not necessarily cross either major line of economic demarcation. Such farmers may be classified as economically mobile, but the precise alteration in social status each experienced is problematic.

For farmers who began the 1850s without slaves, the chances of purchasing bondsmen during the coming decade were not bright (see Table 2.3). As will become clear, in every region farmers at the bottom of the economic pyramid were far less likely to acquire slaves than to acquire land. Above all, this pattern reflected the high price of slaves relative to land and the indivisibility of slave property as a form of capital.[45] Pro-

43. Compare persistence rates given in Table 2.2 with the following for farm operators in the Midwest: Trempeleau County, Wisconsin, 1860–70, 31.9 percent; Grant County, Wisconsin, 1880–95, 22 percent; Eastern Kansas, 1870–80, 44.1 percent; Crawford Township, Iowa, 1850–60, 31.8 percent; Wapello County, Iowa, 1850–60, 41.4 percent; Bureau County, Illinois, 1850–60, 37.7 percent. (Statistics for Wapello and Bureau counties apply to both farm operators and farmers without farms.) See Curti, The Making of an American Community, p. 70; Peter J. Coleman, "Restless Grant County: Americans on the Move," *Wisconsin Magazine of History* 46 (1962):16–20; James C. Malin, "The Turnover of Farm Population in Kansas," *Kansas Historical Quarterly* 4 (1935):365; Allan Bogue, *From Prairie to Corn Belt*, p. 26; Mildred Throne, "A Population Study of an Iowa County in 1850," *Iowa Journal of History* 57 (1959):311.

44. Wright, *The Political Economy of the Cotton South*, p. 37.

45. Donald Schaefer, "Yeoman Farmers and Economic Democracy," p. 435.

Table 2.3 Mobility into and out of the slaveholding ranks among heads of free farm households, sample Tennessee counties, 1850-1860 (persisters only)

	East	Middle	West
Percentage of 1850 slaveless owning slaves in 1860	8.2	22.4	29.9
Percentage of 1850 slaveholders owning no slaves in 1860	28.4	23.6	12.5

Source: Eight sample counties. See text.

spective landowners in any area of the state could purchase small farms for far less than the cost of a single adult slave.[46] Accordingly, in all sections instances of slaveless farmers amassing large numbers of bondsmen were exceedingly rare. The median number of slaves owned by new slaveowners in 1860 was two in East and Middle Tennessee and only three in West Tennessee.

The experience of Tennesseans who began the decade as masters shows the tenuous nature of slaveownership, especially among smaller slaveowners. The likelihood of a slaveowner losing all his slaves was particularly great in Middle and East Tennessee, where nearly one-fourth or more of persisting slaveowners were slaveless by 1860. With rare exceptions the downwardly mobile in all three sections had been small slaveholders; more than half owned three or fewer slaves. The data strongly confirm James Oakes's opinion that "the majority of slaveholders who owned no more than five bondsmen were not a stable economic class."[47]

The acquisition of land was a far more common form of upward mobility than was the acquisition of slaves. In each region at least two-fifths of landless farm household heads who persisted throughout the 1850s purchased land sometime during the decade (see Table 2.4). Unexpectedly, the success rate was significantly greater in the plantation counties of West Tennessee than in either the mixed-farming area of central Tennessee or

46. The average assessed value of an acre of land in 1859 was $5.14 in East Tennessee, $10.89 in Middle Tennessee, and $8.07 in West Tennessee. In the same year average assessed values per slave, regardless of age or sex, were $800.22, $822.13, and $914.50, respectively. *See Appendix to Senate and House Journals, Tennessee, 1859–1860* (Nashville: E. G. Eastman & Co., 1860), pp. 26–31.

47. James Oakes, *The Ruling Race: A History of American Slaveholders* (New York: Random House, 1982), p. 40.

Table 2.4 Mobility into and out of the landowning ranks among heads of free farm households, sample Tennessee counties, 1850-1860 (persisters only)

	East	Middle	West
Percentage of 1850 landless owning land in 1860	44.4	53.8	65.1
Percentage of 1850 landowners owning no land in 1860	6.6	6.3	7.8

Source: Eight sample counties. See text.

the yeomen-dominated eastern reaches of the state. Overall, the proportion of persisting tenants and farm laborers gaining farms of their own in West Tennessee was 20 percent greater than in Middle Tennessee and nearly 50 percent greater than in eastern Tennessee.[48] In every section, however, the number of successful climbers was sufficiently large to suggest that an "agricultural ladder" was clearly operative within the farm population.

Movement along this ladder went in two directions, of course. Although landless farmers who persisted commonly acquired farms of their own during the 1850s, a small but significant percentage of landowning farmers lost theirs. Unlike the proportion of landless farmers who acquired land, the proportion of landowners who lost their farms varied negligibly from region to region. Because the rate of outmigration was probably greater among farmers who lost their land than among the farm population as a whole, Table 2.4 probably underestimates the proportion of farmowners who lost their land. Still, relatively few farmers would look back on the decade as a period of decline into landlessness. Many, on the other hand, would remember those years fondly as a time of upward mobility.

Occasionally, the extent of such elevation was spectacular. Across the state between 6 and 9 percent of persistent farmers who began the decade without land had risen to a position among the top quintile of local land-

48. However, by calculating rates of land acquisition with reference to the total number of landless households in 1850, rather than to persisting households only, an entirely different conclusion is indicated. The proportion of landless farmers in 1850 who persisted throughout the decade and succeeded in acquiring land was actually lowest in West Tennessee, approximately 15 percent. It was significantly higher in Middle Tennessee (20 percent) and in East Tennessee slightly higher still (21 percent). Clearly, no solid judgment concerning variations in economic opportunity is possible without greater understanding of the social and economic significance of migration in the local context.

owners by the decade's end. John P. Snapp of Greene County, for example, was a tenant in 1850 but by 1860 had acquired more than 2,300 acres of land and fifteen slaves. Four hundred miles away William B. Taylor of Haywood County rose from tenancy to ownership of a plantation worth more than $70,000 during the same span of time. Such meteoric mobility was rare, however, and, if Snapp and Taylor were representative, it was often facilitated by inheritance from elite relatives.[49]

In all sections of the state the typical success story during the 1850s was far more modest. In each area, between two-thirds and three-fourths of landless farmers who acquired land during the decade remained in the bottom half of local landowners in 1860 in terms of the value of their holdings.[50] If "a lot of poor men got rich," as Middle Tennessean James Williams claimed, then with few exceptions they did so after decades, not years, of struggle. Williams's viewpoint was undoubtedly influenced by the success of his father, who "started out with no property" but had accumulated by 1860 "a nice estate" of 1,200 to 1,400 acres of land and approximately fifty slaves. Veterans who had witnessed greater struggle were apt to recall opportunities for advancement differently. As East Tennessean Joel Acuff recalled, "it was a hard task for a young man to buy a farm or start into business." "Conditions were such," a Wilson County veteran agreed, "that it took a long time to accumulate enough money." Even so, for antebellum farmers who were willing to wait, the prospects of future, if not immediate, landownership were promising regardless of region (see Figure 2.3). The fraction of farm household heads who owned land rose steadily by age cohort and reached extremely high proportions toward the end of the life cycle.[51]

This impression of modest but widespread mobility is reinforced by an examination of *relative* mobility, that is, by broadening the focus to examine the experience of all persisting farmers, including those who both began and ended the 1850s as landowners. Such farmers invariably reported more valuable real estate in 1860 than in 1850, a result of enlarged holdings, appreciation due to capital improvements, or the general increase of land values during the period. On average the value of persisting landowners'

49. Tennessee State Library and Archives, Greene County Records, Office of the Registrar of Deeds, volume 25, pp. 176–8; Haywood County Records, Office of the Registrar of Deeds, book R, p. 43, book T, p. 533.
50. The proportion was 73.6 percent in East Tennessee, 68.2 in Middle Tennessee, and 62.4 in West Tennessee.
51. Dyer and Moore, comps., *Tennessee Veterans Questionnaires*, vol. 5, pp. 2198–9; vol. 1, p. 1; vol. 5, p. 2258.

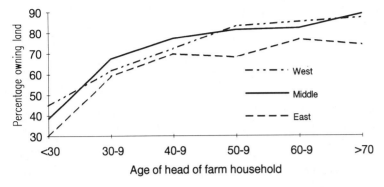

Figure 2.3   Landownership by age, sample Tennessee counties, 1860

real estate increased by 67 percent in East Tennessee, 175 percent in Middle Tennessee, and 137 percent in West Tennessee.[52]

Appreciation in the value of a farmer's property did not necessarily imply change in social status, however. In contrast to the assessment of farmers who moved from landlessness to landownership or vice versa, it is much more difficult to determine when the wealth position of a persistent landowner had changed sufficiently to qualify him as upwardly or downwardly mobile. Clearly, Haywood County farmer John R. Watkins, whose real wealth increased from $2,500 to $30,000 during the 1850s, was an individual who had moved from one economic position to another in his community. But what about Johnson County's Thomas Dougherty, whose real property increased in value from $170 in 1850 to $400 in 1860? Did he experience a qualitative shift in hierarchical position? Although he may have perceived improvement in his condition, one may question whether the absolute change in Dougherty's wealth was large enough to imply a perceived change in status.

It is precisely because of such ambiguity that arbitrary definitions of absolute economic mobility among landowners are impractical. It is far better simply to measure the extent of relative movement within the landholding ranks, of whatever magnitude, that characterized the farm population in each region. Toward this end, persisting farmowners have been assigned decile rankings for 1850 and 1860 that are based on the value of their real property in relation to the overall distribution of real estate among

52. Based on the median value among individual persisting landowners of the ratio of 1860 real wealth to 1850 real wealth. The proportion of persisting landowners reporting a decline in value of real wealth was 2.7 percent in West Tennessee, 3.4 percent in Middle Tennessee, and 5.2 percent in East Tennessee.

Table 2.5 Change in real wealth decile rank among heads of free farm households, sample Tennessee counties, 1850-1860 (persisters only)

	East	Middle	West
Percentage moving upward	44.3	48.7	40.5
Percentage moving downward	18.9	22.1	29.2
Percentage stable	36.9	29.2	30.3
Mean absolute decile change	1.7	1.8	1.9

Source: Eight sample counties. See text.

landowners within their counties. Farmers who were landless in either or both years received decile ranks of zero for the appropriate year. Although far from perfect, the average change in decile level among persisters may be interpreted as a crude measure of the degree of fluidity along the continuum between landless tenant or laborer and wealthy landholder.

In every region of the state, between two-fifths and one-half of persisting farm household heads moved up in decile position during the prosperous decade of the 1850s (see Table 2.5). Upward movement was most common in the mixed-farming communities of Middle Tennessee and least prevalent within the Black Belt counties to the west. At the same time, western farmers were more likely to fall in decile position than were farmers elsewhere. Even so, the differences among the sections were not great, and substantially similar patterns obtained in all three. Indeed, the overall extent of movement along the real wealth continuum, as indicated by the mean absolute decile change per persister, was almost identical among the three regions.

The statistics given in Table 2.5 exaggerate fluctuations among persisting farmers because they group farmers who moved only one decile with those who moved ten. A more detailed inspection of relative mobility patterns reveals that in every region a striking degree of stability characterized the entire landholding continuum (see Table 2.6). Farmers who began the decade at or near the bottom of the local continuum were likely to move the furthest – upward one to two deciles on average. Individuals with more substantial holdings (i.e., those ranked among the top seven deciles of landowners in 1850) averaged moving one decile or less. All told, even though real estate values boomed during the 1850s, one-half to two-thirds of farmers died or moved away, and hundreds of newcomers moved in to take

Table 2.6 Movement of persisting free farmers in real wealth position relative to local free farm population, sample Tennessee counties, 1850-1860 (heads of household only)

	East	Middle	West
Real wealth decile in 1850	*Median real wealth decile in 1860*		
0	0	1	2
1	2	3	3
2	3	3	3
3	5	4	4
4	5	5	4
5	6	6	5
6	7	7	6
7	7	7	6
8	8	7	7
9	9	9	9
10	10	10	10
Percentage moving 0-1 deciles	60	59	61

Source: Eight sample counties. See text.
Note: Farmers with "0" decile rank in 1850 were landless. Farmers with "1" rank were in poorest decile of landowners; farmers with rank of "10" were in the wealthiest.

their places, in every section fully three-fifths of farmers who persisted either moved one decile only (up or down) during the decade or experienced no change whatsoever.

Whether this extent of stability was exceptional within the American context is impossible to determine at present. There are so few comparable studies of economic mobility in other rural areas, either within or outside of the South, that the frame of reference necessary for comparison is lacking.[53] One unassailable conclusion emerges, however: Among the three regions studied here, patterns of relative economic mobility were remarkably similar. It evidently mattered little whether a farmer lived in the Appalachian foothills of East Tennessee, the prosperous, commercially oriented breadbasket of the Central Basin, or the slave and cotton kingdom

53. Kearl and Pope provide excellent, detailed data on economic mobility in mid-nineteenth-century Utah, but it is questionable whether patterns observed for that frontier territory closely resembled those for more settled agricultural regions. See "Wealth Mobility: The Missing Element."

near the banks of the Mississippi. Whatever their location, farmers typically moved slightly if at all in economic position relative to their neighbors. Although these patterns are testimony to the impressive stability of the socioeconomic order in each region, they do not imply stagnation in the lives of individual farmers. To gauge the extent of *lifetime* mobility it is necessary to extend the trends of the 1850s across several decades. Not anticipating the cataclysm of civil war, farmers who moved upward slightly during the final antebellum decade had reason to expect noticeable improvement over the course of their adult lives. Slow, modest upward mobility, as defined within the local context, was widespread among persisting farmers in every area studied.

Conclusion

Although one can only applaud scholars' increasing sensitivity to the Old South's internal diversity, the results of this analysis of wealth concentration and economic mobility in antebellum Tennessee indicate that recent scholarship may actually exaggerate southern heterogeneity. Despite large internal variations in agricultural emphasis and in dependence on slavery, in Tennessee fundamental structural similarities unified plantation and nonplantation regions. In particular, despite great variation across the state in the magnitude and composition of wealth, local patterns of wealth distribution were literally indistinguishable from the Appalachians to the Mississippi. A highly skewed concentration of wealth was not peculiar to the Black Belt but a prominent feature of each of the state's major sections, including such yeomen-dominated areas as the East Tennessee Valley. Nor were there significant distinctions across the state in the pace at which wealth was acquired. In every region of the state a highly uneven distribution of wealth coexisted with a social order sufficiently fluid to sustain the egalitarian ideal that hard work should bear fruit in economic independence. As Sims Latta, the son of a Tennessee tenant farmer, recalled about the prewar years, "it took some time to accumulate but honest industry in time brought good recompense."[54] Although referring specifically to his own community, Latta's observation held true across a state renowned for the diversity of its land and its people.

54. Dyer and Moore, comps., *Tennessee Veterans Questionnaires*, vol. 4, p. 1336.

3. "God Only Knows What Will Result from This War": Wealth Patterns among White Farmers, 1860–1880

"The political affairs of the nation disturb my mind," West Tennessee planter John Alexander Taylor confided to his diary in mid-February 1861. Like many large slaveholders, Taylor had mixed feelings during the secession winter of 1860–61. Despite his ambivalence, or possibly because of it, Taylor ultimately sided with the vast majority of his Haywood County neighbors in the state referendum in June, casting his ballot for separation from the Union and representation in the Confederate Congress. He did so, however, with neither animosity nor secessionist zeal. "Better to make our own laws," he noted in simple justification of his act, "than to be always quarreling."[1]

Taylor's terse diary entries do not reveal the cause of his initial hesitation. Perhaps, as was true of numerous southern moderates – the prescient

---

1. Taylor's diary is reprinted in Taylor Kinfolk Association, *The Taylors of Tabernacle: The History of a Family* (Brownsville, TN: Tabernacle Historical Committee, 1957), quotations from p. 177. Although 93 percent of Fayette and Haywood county voters endorsed separation in the June 8 referendum, in the presidential election the preceding November they had roundly rejected the extreme southern rights candidate, John C. Breckinridge. John Bell and Stephen Douglas, both compromise candidates, combined for 80 percent of the popular vote. Concerning popular pressures to support secession in Haywood County, see the testimony of Taylor's cousin in "Testimony of Henry L. Taylor before the Commissioners of Claims," Records of the U. S. House of Representatives, 1871–1880, RG 233, National Archives and Records Administration [hereafter cited as NARA]. For county-level voting returns, see Mary E. R. Campbell, *The Attitude of Tennesseans Toward the Union, 1847–1861* (New York: Vantage Press, 1961), pp. 284–94. On the conversion of southern unionists during the secession crisis, with particular emphasis on Tennessee, see Daniel W. Crofts, *Reluctant Confederates: Upper South Unionists in the Secession Crisis* (Chapel Hill: University of North Carolina Press, 1989), pp. 334–52.

ones – he recognized secession's inherent threat to the stability and security of his world.[2] Without question, there was much to be lost should the gamble fail: The Taylors were one of the oldest and wealthiest families in the county. Four days away from his forty-third birthday when Tennessee left the union, John Taylor had lived in Haywood since the age of six, when in 1825 his father, Richard, had moved his wife and three sons from Mecklenburg County, Virginia, to the cotton frontier. An extensive migration of kinfolk followed, and by 1832 his grandfather and four uncles had also built homes in the same neighborhood east of Brownsville. The family flourished over the years, and by 1860 they had accumulated collectively nearly 500 slaves and 20,000 acres of land. John Taylor had also succeeded individually; when war began he owned some fifty-six slaves and a plantation of 1,200 acres adjacent to his father's.[3]

Nearly a year would pass before the war impinged directly on Taylor's operations. "Times are getting hot with the war," he observed on the eve of Bull Run, yet the daily routine of plantation life was ostensibly unchanged throughout 1861. Certainly there were drawbacks – "rain[,] War, and complaining negroes" were sources of discouragement – yet the land still yielded 110 bales of cotton in the fall, nearly double the output of just the year before. In February 1862, however, the fall of Fort Henry to Ulysses Grant transferred control of the Tennessee River to the Union, rendered West Tennessee vulnerable to Federal invasion, and set in motion a process that altered Taylor's way of life irrevocably.[4]

Ironically, much of the initial disruption of his plantation stemmed from Confederate attempts to defend the region. In March Confederate authorities impressed three of his slaves to work on defenses at Island Number Ten, a strategic stronghold in the Mississippi River north of Memphis. In late May came the order to burn all cotton in the neighborhood to prevent its confiscation by Union invaders. Before their final withdrawal from the region in early June, Confederate troops completed their work of destruction, burning bridges over both major rivers in the county, the Hatchie and the Forked Deer. Nine days later Federal troops occupied Brownsville. "It is generally thought the South is about to be subjugated," Taylor re-

2. On the diversity of sentiment among planters during the secession crisis, see James L. Roark, *Masters Without Slaves: Southern Planters in the Civil War and Reconstruction* (New York: W. W. Norton & Company, 1977), pp. 1–32.
3. *The Taylors of Tabernacle*, pp. 3, 15.
4. Ibid., diary entries for 20 July, 29 August, and 13 December 1861.

corded in his diary in anticipation of the Confederate evacuation. "God only knows what will result from this war."[5]

Given without benefit of hindsight or detachment, such anguished cries in the face of the unknown reverberated across the war-torn Confederacy. Long after Appomattox, white Southerners continued to be unsure concerning the ultimate meaning of the war for themselves and their society. Confusion remained the watchword among individuals who knew for certain only that history had happened to them and that it had been painful. Journalists, philosophers, and politicians aside, most whites probably were too busy reacting to the war's short-term implications – poverty, hunger, and loss of loved ones – to spend much time contemplating the long-term implications of the conflict. Historians have assumed the latter task with a vengeance, however, writing with a dark fascination about the death throes of the Confederacy and the effects of military defeat and emancipation on the economy and society of the region.[6] As with so many issues in southern history, however, the great diversity of the region complicates the question of how or whether the "new" South was new. Unfortunately, scholarly assertions concerning the extent of southern continuity or discontinuity frequently rest on facile generalizations that minimize or ignore both antebellum and postbellum heterogeneity. Sensitive to the diversity of the nineteenth-century South, both before and after the Civil War, this chapter reexamines the issue of southern discontinuity through an analysis of the sample counties between 1860 and 1880. Beginning with the decade of war and emancipation, it investigates changes in the level and distribution of wealth and in patterns of migration and socioeconomic mobility. Although source limitations preclude an equally detailed analysis of the 1870s, the discussion offers tentative conclusions concerning that decade as well.

War, Emancipation, and the Decline in Wealth, 1860–1870

As in other Confederate states, the Civil War wrought unprecedented physical destruction and economic disruption in the state of Tennessee. In 1872

5. Ibid., diary entries for 7 March, 28 May, 5 June, and 29 April 1862.
6. A good recent review of the literature is Harold Woodman, "Economic Reconstruction and the Rise of the New South, 1865–1900," in John B. Boles and Evelyn Thomas Nolen, eds., *Interpreting Southern History: Historiographical Essays in Honor of Sanford W. Higginbotham* (Baton Rouge and London: Louisiana State University Press, 1987), pp. 254–307.

a congressional committee estimated the state's losses, including both slave and non-slave property, at over $185 million, a staggering 2.5 times the gross agricultural product of the state during the banner year of 1859.[7] Although no corner of Tennessee was unaffected, the costs of the war clearly were not distributed uniformly across the state. Just over half of the financial loss, $96.5 million, stemmed directly from the emancipation of the state's 275,000 slaves. This burden would have been relatively light in the sample East Tennessee counties, where slaveowners constituted less than one-tenth of farm households, far heavier in those in Middle Tennessee, where one-third of farmers had owned slaves, and heaviest of all in the sample southwestern counties, where two-thirds of farmers had been masters.

In addition to the obvious disappearance of slave wealth, the extent of material destruction in the state was also appalling. Congress placed the total value of nonslave property losses in the state at approximately $89 million, or just under one-third of the total value of nonslave property when the war began.[8] All across the state, Confederate and Union veterans returned from the war to find dilapidated buildings, deteriorated soils, broken-down work stock, and depleted food supplies. When Joel Henry returned to his home in East Tennessee, for instance, he found that "conditions were awful – fences burned, provisions about all taken – stock driven off, houses burned – nothing to make a crop on." A veteran from Middle Tennessee recalled that when he departed for the army in 1861 he left behind "$5000 in cash and good notes . . . seven head of horses four

---

7. Using county-level aggregate data on crop outputs and livestock inventories, and applying wholesale price statistics for regional market centers (Knoxville, Nashville, and Memphis), I estimate the value of gross agricultural output for the state in 1859 at approximately $73.1 million. (For wholesale price figures, see Appendix B.) For figures on estimated property losses, see "Affairs in the Late Insurrectionary States," H. R. 22, 42d Cong., 2d sess., 1872, Pt. 1., p. 110. The congressional figures, particularly those pertaining to slave losses, are unquestionably gross underestimates. On the costs of the war, see James L. Sellers, "The Economic Incidence of the Civil War in the South," *Mississippi Valley Historical Review* 14 (1927–28):179–91; and Claudia Goldin and Frank D. Lewis, "The Economic Cost of the American Civil War: Estimates and Implications," *Journal of Economic History* 35 (1975):299–326. Roger Ransom and Richard Sutch maintain that historians have exaggerated the extent of physical destruction in the South and argue that the voluntary withdrawal of black labor after emancipation was far more important than the loss of work stock or implements. See "The Impact of the Civil War and of Emancipation on Southern Agriculture," *Explorations in Economic History* 12 (1975):1–28.
8. "Affairs in the Late Insurrectionary States," p. 110.

cows ten hogs. Wife and two babys. When I got back had wife two babys [and] a sack of Confederate money. . . . [N]o bread or meet for wife and babys." Thomas Corn, a veteran from a nearby county, recounted a similar experience. Corn returned to his home in May 1865 to face the "hardest strugle" of his life. "[T]heare was nothing . . . on the farm to live on and nothing to eat. [N]ot a hog cow calf hores or mule and not a chicken."[9]

The extent of nonhuman property damage was more evenly distributed than the losses associated with emancipation, yet there were still significant variations across the state. Perhaps the greatest irony of Tennessee's Civil War was that the area most forcefully opposed to secession was made to endure Confederate rule until very nearly the end of the war, whereas the sections where secessionist fervor had been strongest fell quickly to Union occupation. Consequently, the experience of the regions differed significantly, albeit less in the extent of physical devastation than in the circumstances under which it was inflicted.

Confederate and Union armies struggled for control of the Valley of East Tennessee intermittently for three years, marching almost the whole length of the region four separate times and waging major campaigns for the area in 1863 and 1864. The valley was not permanently under Federal control until late 1864, and it was early 1865 before Union forces established dominance in the remote Appalachian counties northeast of Knoxville. In the meantime, both armies supplied themselves by living off the land. The Confederate retreat after the failed siege of Knoxville in late 1863 was particularly destructive; believing that they were leaving the region for good, the Confederates took everything they could with them. A Union officer stationed in the region at the time reported of "the ravages of both Union and rebel armies and the extreme destitution produced by them among the people." During the winter of 1864 Confederate general James Longstreet, who had commented only the previous year concerning the great abundance of the region, wrote to Robert E. Lee from his winter quarters near Greeneville that "there is nothing between this and Knoxville [a distance of sixty miles] to feed man or horse."[10]

9. Gustavus W. Dyer and John Trotwood Moore, comps., *The Tennessee Civil War Veterans Questionnaires* (Easley, SC: Southern Historical Press, 1985), vol. 3, p. 1081; vol. 4, pp. 1511–12; vol. 2, pp. 562–3.

10. James B. Campbell, "East Tennessee During the Federal Occupation, 1863–1865," *East Tennessee Historical Society's Publications* 19 (1947):64–80 (quotation on p. 70); J. E. Jacobs to Clinton B. Fisk, 28 May 1866, in "Records of the Assistant Commissioner for the State of Tennessee," Bureau of Refugees, Freedmen, and Abandoned Lands, Record Group 105, NARA, microcopy T142; Wilma Dykeman,

The depredations of Federal troops were particularly galling to the predominantly unionist population. Although the region had rejected secession by greater than a 2:1 majority and had sent more than 20,000 volunteers into the Union Army, in terms of physical treatment, at least, East Tennesseans found Union occupation no more desirable than Confederate. Prominent loyalist politician T. A. R. Nelson observed as much in a letter to a Union general at the end of 1863:

> The Union Army is more destructive to Union men than the rebel army ever was. Our fences are burned, our horses are taken, our people are stripped in many instances of the last vestiges of subsistence, our means to make a crop next year are being rapidly destroyed, and when the best Union men in the country make appeals to the soldiers, they are heartlessly cursed as rebels.

As a result, Nelson maintained, loyal citizens were coming to feel that the government that they had "loved and honored and trusted . . . had at last become cruel and unjust, and care[d] nothing for [their] sorrows and sufferings." Although the conclusion was understandable, and at least partially justified, much of the damage inflicted by Union soldiers stemmed from necessity more than insensitivity. As Union general Gordon Granger observed, although nothing had troubled him "so much as being compelled to strip the country[,] friend and foe must fare alike, or the army must starve."[11]

The French Broad (New York: Rinehart & Company, Inc., 1955), p. 116. See also Thomas W. Humes, *Report to the East Tennessee Relief Association at Knoxville* (Knoxville: n.p., 1865); idem, *The Loyal Mountaineers of Tennessee* (Knoxville: Ogden Bros., 1888); Spencer Henry to Andrew Johnson, 20 September 1864, and Jacob M. Bewley to Andrew Johnson, 2 December 1864, in Leroy P. Graf and Ralph W. Haskins, eds., *The Papers of Andrew Johnson*, vol. 6 (Knoxville: University of Tennessee Press, 1983), pp. 176, 325–6. Charles Faulkner Bryan Jr. provides the best scholarly discussion of military destruction in East Tennessee in "The Civil War in East Tennessee: A Social, Political, and Economic Study," (Ph.D. diss., University of Tennessee, 1978); but see also Harold S. Fink, "The East Tennessee Campaign and the Battle of Knoxville in 1863," *East Tennessee Historical Society's Publications* 29 (1957):79–117.

11. Thomas A. R. Nelson to Brig. General S. P. Carter, 26 December 1863, in *War of the Rebellion: A Compilation of the Official Records of the Union and Confederate Armies* (73 vols., Washington, DC: Government Printing Office, 1880–1901), XXXI, pt. 3, p. 508 (hereafter cited as *OR*); Dykeman, *The French Broad*, p. 112. See also General U. S. Grant to General W. T. Sherman, 1 December 1863, *OR* XXXI, pt. 3, p. 297; Sherman to Grant, 11 December 1863, *OR* XXXI, pt. 3, p. 382; John Netherland to Andrew Johnson, 7 August 1864, in Graf and Haskins, *Papers of Andrew Johnson*, vol. 6, pp. 26–7; Greeneville *New Era*, 4 November 1865; Brownlow's *Knoxville Whig and Rebel Ventilator*, 29 March 1865.

In contrast to East Tennessee, most of the remainder of the state – where voters had favored secession by a 6:1 margin – came under Federal military rule during the first half of 1862. Union troops occupied Nashville and the rest of upper Middle Tennessee (including Robertson and Wilson counties) in late February 1862, and Memphis and much of western Tennessee (including Haywood and Fayette counties) fell to Union forces less than four months later. Although small Confederate units continued to create havoc in these areas, and isolated communities often changed hands more than once, both sections remained more or less under Union control for the duration of the war.[12]

For the most part, as a result, these areas escaped the extreme physical destruction produced by major military confrontations. Indeed, one historian of Haywood County has argued that its loss "in the way of property was perhaps as light as that of any other county in the state." The depredations accompanying three years of military occupation were substantial, however. As in East Tennessee, it mattered little whether the color of the uniforms was blue or gray – the proximity of soldiers invariably resulted in widespread destruction of livestock and crops. A West Tennessee landowner lamented that soldiers of both armies took turns stealing his property, "each pretending to fear it may fall into the hands of the other and be turned against them."[13]

As one of the leading food-producing regions of the Confederacy, Middle Tennessee seems to have suffered especially under Union occupation. "It is really sad to see this beautiful country ruined," an Indiana cavalryman stationed near Nashville observed during the summer of 1863. "There is no corn and hay for the cattle and horses, but there are no horses left anyhow and the planters have no food for themselves. . . . I can't see how

12. Southern Middle Tennessee (including Lincoln County) was also under federal control, except for the period between the fall of 1862 and the summer of 1863 and a few weeks late in 1864 before the battles of Franklin and Nashville. See Stephen V. Ash, *Middle Tennessee Society Transformed, 1860–1870: War and Peace in the Upper South* (Baton Rouge: Louisiana State University Press, 1988), p. 85; and Joseph H. Parks, "A Confederate Trade Center under Federal Occupation: Memphis, 1862–1865," *Journal of Southern History* 7 (1941):289–314 (quotation on p. 291). The proportion of voters supporting secession was 88 percent in Middle Tennessee and 83 percent in West Tennessee. See Campbell, *The Attitude of Tennesseans Toward the Union*, pp. 292–4.

13. *Goodspeed History of Tennessee* (Nashville: The Goodspeed Publishing Company, 1887), Pt. 2, p. 825; John Houston Bills Diary, Southern Historical Collection (microfilm copy in Tennessee State Library and Archives), entry for 19 October 1863.

the people live at all." Describing to his wife how the army sustained itself, he explained that "when we have eaten a place empty, we go a few miles farther and take everything there we can find." Union soldiers in Middle Tennessee had made foraging into a "military science," an officer in the Army of the Cumberland noted jocularly in 1864. "Woe unto pigs and sheep and calves and chickens when they are on the march!" Mindful of the human cost, a Union general stationed in the region criticized the foraging system practiced by the Federal troops: "all suffer, rich and poor; of all methods of providing for any army this is the most wasteful."[14]

The situation in West Tennessee was similar, if marginally less severe. The ordeal of Franklin D. Cossitt, a wealthy LaGrange cotton planter and merchant, was revealing, if undoubtedly extreme. Before evacuating Fayette County in late May 1862, Confederate agents forced Cossitt to lead them to several caches of cotton that the planter had stockpiled on his three plantations. They then burned the cotton – a total of "two hundred and sixty odd bales," worth $20,000 in gold if a penny – to prevent its confiscation by the approaching Federals. Not long thereafter a Union army of 10,000 men camped on his largest plantation, remaining there for several weeks. "After they finally left," recalled Cossitt, a Connecticut-born unionist, "I found everything gone – it was a general cleaning up; it was stripped of everything."[15]

14. Fanny J. Anderson, ed., "Documents: The Shelly Papers," *Indiana Magazine of History* 44 (1948):187, 193; John Fitch, *Annals of the Army of the Cumberland* (Philadelphia: J. B. Lippincott & Co., 1864), p. 655; John M. Palmer, *Personal Recollections of John M. Palmer* (Cincinnati: The R. Clarke Co., 1901), pp. 137–8. See also David M. Smith, ed., "Documents: The Civil War Diary of Colonel John Henry Smith," *Iowa Journal Of History* 47 (1949):142; A. T. Volwiler, ed., "Documents: Letters from a Civil War Officer," *Mississippi Valley Historical Review* 14 (1928):512; Ash, *Middle Tennessee Society Transformed*, pp. 85–92; and Peter Maslowski, *Treason Must Be Made Odious: Military Occupation and Wartime Reconstruction in Nashville, Tennessee, 1862–1865* (Millwood, NY: KTO Press, 1978), pp. 131–6.

15. "Testimony of Franklin D. Cossitt before the Commissioners of Claims," in Records of the General Accounting Office, claim #20459, Record Group 217, NARA. For other accounts of Confederate destruction of cotton, see idem, "Testimony of Thomas Bond Before the Commissioners of Claims," claim #19890; "Testimony of Gray B. Medlin," Records of the U. S. House of Representatives, RG 233, NARA, claim #8660. Cossitt estimated that the Union army confiscated nearly $32,000 worth of nonslave property between the summer of 1862 and the spring of 1863, including 45 mules, 500 hogs, 83 head of cattle, 1,350 bushels of potatoes, and nearly 14,000 bushels of corn. He was ultimately awarded a little over $14,000 by the federal government some fourteen years after the property was taken. A general discussion of the claim procedure may be found in Frank W. Klingberg, *The Southern Claims Commission* (Berkeley: University of California Press, 1955).

In contrast, Union and Confederate soldiers bled John A. Taylor dry a little at a time. The first Federal troops to visit Taylor's plantation arrived in September 1862 – "a very wicked, swearing set of men they seem[ed] to be" – and paid their respects by taking four of his best horses. During the course of that fall they gradually took most of his hogs as well. In early 1864 northern soldiers satisfied other tastes, stealing his wife's jewelry as well as $500 in gold from his father. Less than six weeks later Confederate troops temporarily reclaimed his neighborhood from the Yankee thieves but confiscated huge quantities of bacon and corn meal as their reward. If Taylor was grateful, he failed to reveal that in his diary, observing instead that "we are entirely likely to be eaten out by the soldiers."[16]

The diary entries of Hardeman County's John Houston Bills are also illuminating, for they not only detail the incremental destruction that was common in West Tennessee but also highlight the commercial possibilities ⁺hat Union occupation brought to the region. For Bills "the War was distant & consequently only to be heard of and talked about" until the Confederate evacuation of the region in May 1862. "Since then," however, "the war has been upon us at our doors, the horrors of which cannot be written." During the first week of June a Union army under General Lew Wallace occupied Bolivar, the county seat situated a few miles from Bills's plantation; within a few weeks the plundering was well underway. In mid-July Bills noted that Union soldiers were pilfering his crops – "fruits and fowls [were] all going as well." In August he estimated his losses to that point at $100,000 (including departed slaves), but his misfortune had only begun. In September soldiers raided his corn crib and trampled much of his crops still in the field. Subsequent weeks brought nearly daily complaints of Union soldiers who stole his hogs and sheep and harassed his female slaves. By mid-October Union troops had killed almost all his livestock, carried off 20,000 bricks, and stolen one-half of the cotton on his place. At the end of the month they destroyed his cotton gin and gin house; early the following month they burned down his stables.[17]

Although his losses were real and his anguish undeniable, Bills was able to offset partially the ruinous costs of the war by taking advantage of lucrative trade opportunities stemming from the continued northern demand for cotton. Only four months after the fall of Fort Sumter, President Lincoln had authorized commerce with all Confederate areas under Union

16. *The Taylors of Tabernacle*, pp. 185–93 (quotation on p. 193).
17. John Houston Bills Diary, entries for 29 August; 5 June; 14 July; 26 August; 6 September; 2, 13, 29 October; 6 November 1862.

occupation. Consequently, the capture of Memphis the following June not only reclaimed for the Union the largest inland cotton center in the South but also opened up a potentially vast and lucrative trade between farmers in the surrounding countryside and the literal horde of northern merchants who descended on the city. Indeed, farmers could sell their cotton without actually traveling to Memphis, for enterprising buyers authorized to trade across Union lines roamed the hinterlands in search of the valuable crop. Supplies were plentiful in Bills's neighborhood, for the Union Army had entered Hardeman County before retreating Confederates could destroy more than a few bales. By the first of July, cotton was selling in Bolivar at fifteen to eighteen cents (in gold) per pound – 50 to 80 percent above prewar prices – and farmers were "hauling [it] into town as though the war was over." A month later, after prices in the region had topped twenty cents per pound, Bills sold his own stockpile for over $15,000 – far more than he had ever made during the prosperous 1850s. "The high price of cotton," he noted in an understatement, "in some degree helps us in our losses by the War."[18]

The frenzied northern demand for cotton, coupled with the lenient trade policies of Union armies, presented West Tennessee farmers with far greater commercial opportunities than farmers elsewhere in the state, although the demise of slavery ultimately hindered them from exploiting such opportunities fully. Despite the Confederate Congress's prohibition of cotton shipments to any port or trade center under Federal control, tens of thousands of bales were channeled upriver through Memphis. Much of it came from unionists such as Franklin Cossitt or highly reluctant secessionists such as John Houston Bills, men who could send their cotton to northern factories without hesitation or violation of principle. Without question, much of it also came from supporters of the rebellion who overcame their scruples in exchange for high profits. John Alexander Taylor was a good example of the latter. Although hesitant initially in his support of disunion, once Tennessee seceded Taylor quickly became fervent in his support of "the Cause." A close rereading of the book of Ezekiel convinced him that the South was "God's Restored Israel," and he remained confident of Confederate success until the very end. Old Testament prophesy aside, however, Taylor was not deterred from selling his stored-up cotton in August 1862 for twenty-two cents per pound in northern gold. This combination of ideological devotion and economic expediency was not un-

18. Ibid., entries for 1 July, 26 August 1862; Parks, "A Confederate Trade Center under Federal Occupation," pp. 289–301.

usual in Haywood County; when a change in Federal policy restricted the cotton trade to "loyal" participants, 181 of Taylor's neighbors signed oaths of loyalty to the United States within the space of three days.[19]

Despite such opportunism, in the end the resilient cotton trade cushioned but slightly the devastating impact of emancipation on the area. The windfall enjoyed by West Tennessee farmers was generally short-lived. While cotton prices soared ever higher, the rapid disintegration of the peculiar institution after the summer of 1862 crippled the productive capacity of the region far more than the combined depredations of both armies. Even though Tennessee was officially exempted from the Emancipation Proclamation, the prolonged and extensive presence of Union troops fundamentally undermined the master–slave relationship long before the slaves were legally freed. Union army recruiters conscripted thousands of West Tennessee slaves into military service, and thousands more found "freedom" in Federal "contraband" camps. If the impressions of numerous masters are reliable, those who remained with their owners labored grudgingly at best. Madison County slaveowner Robert Cartmell, for instance, noted that his slaves had become "entirely corrupted" by the fall of 1862. "Accustomed to keeping the Sons of Ham in their proper place," Cartmell found their "impudence . . . hard to endure." Another West Tennessee master voiced similar complaints less than two months after selling cotton for an unprecedented profit. The slaves were "now worthless for work," he observed with frustration in early October. The result was direct and painfully clear: "Many who thought they were rich are now poor indeed."[20]

Such severe losses inevitably affected wealthholding patterns among white Tennesseans at both the local and state level. The consequences of war and emancipation for *local* wealth distribution were complicated and, as will be shown, defy easy generalizations. At the *state* level, on the other

19. E. Merton Coulter, *The Confederate States of America, 1861–1865* (Baton Rouge: Louisiana State University Press, 1950), p. 286; Parks, "A Confederate Trade Center under Federal Occupation"; *The Taylors of Tabernacle*, pp. 184, 189, 191, 199; B. I. Wiley, "Vicissitudes of Early Reconstruction Farming in the Lower Mississippi Valley," *Journal of Southern History* 3 (1937):441–52; "List & Description of persons who have Subscribed to the Oath of allegiance at Brownsville, TN.," Records of U.S. Continental Commands, Record Group 393, series E157, NARA.

20. Robert H. Cartmell Diary, 13 November 1862, Tennessee State Library and Archives; John Houston Bills Diary, 9 October 1862. For the collapse of slavery in Tennessee during the war, see John Cimprich, *Slavery's End in Tennessee, 1861–1865* (University: University of Alabama Press, 1985); and Bobby L. Lovett, "The Negro's Civil War in Tennessee," *Journal of Negro History* 61 (1976):36–50.

Table 3.1 Median wealth of white farm households, sample Tennessee counties, 1860-1870

	East	Middle	West
Value of real wealth			
1860	$500	$1,500	$1,950
1870	200	700	1,000
Percentage change	-60	-53	-49
Value of personal wealth			
1860	$393	$1,166	$3,500
1870	320	550	400
Percentage change	-19	-53	-89
Value of total wealth			
1860	$900	$3,040	$5,500
1870	600	1,300	1,300
Percentage change	-33	-57	-76

Source: Eight sample counties. See text.

hand, repercussions were obvious and dramatic. The most striking economic effect of Tennessee's Civil War was a drastic reduction of the huge wealth differences that had previously separated plantation and nonplantation regions.

As Table 3.1 shows, emancipation was central to this leveling process. Median real wealth per white farm household declined substantially in all three sections of the state between 1860 and 1870. Although the absolute change differed greatly among the sections, the proportional reduction was far more similar, ranging from just under one half in West Tennessee to three-fifths in East Tennessee, where the drop was actually greatest relative to prewar levels. In each division the decline in real wealth reflected a variety of factors, including increased landlessness, shrinking farm size, soil deterioration, damage to buildings and fences, and falling market values caused by diminished expectations of future profitability.[21] Although the

21. Serious defects in the 1870 agricultural census prevent an accurate comparison of farm values between 1860 and 1870. Tax assessments on real estate constitute a potential alternative but appear highly unreliable. With regard to the eight sample

relative importance of each factor must have varied widely, what is most important in this context is that their aggregate effect was roughly comparable across the state.

In contrast, the losses due to emancipation differed mightily from region to region. Median personal wealth per white farm household declined but 19 percent in East Tennessee, where less than one- tenth of farmers had owned slaves. The proportional reduction was nearly three times as high in Middle Tennessee, where one-third of farmers had been slaveholders, and more than 4.5 times higher in West Tennessee, where two-thirds of farmers had owned an average of ten slaves apiece. In the latter region, average personal wealth plummeted by a staggering 89 percent. Falling levels of total personal wealth, of course, were not precisely equivalent to losses of slave property; the former consisted not only of slaves but of all forms of movable property, including, for example, livestock, household furniture, and jewelry. Even so, the value of human property dwarfed all other categories of personal estate before the Civil War; hence, the figures on personal wealth constitute compelling if imperfect evidence of the financial impact of emancipation.[22]

Almost solely, then, emancipation effected a fundamental *interregional* redistribution of wealth among the state's white farm population. The late antebellum period had been characterized by enormous disparities in overall wealth level among the state's primary sections. Median total wealth per free farm household had been more than three times higher in the state's Central Basin than among the Appalachian counties to the east. It had been far higher still in the major Black Belt counties of West Tennessee, where total wealth per household was nearly double that for Middle Tennessee and more than six times greater than in the eastern part of the state. By 1870, however, due entirely to the differential decline in personal wealth,

counties between 1859 and 1867 (the date of the first complete postwar assessment), the assessed value of land per acre fell approximately 21 percent in East Tennessee, 45 percent in Middle Tennessee, and 17 percent in West Tennessee. The accuracy of these figures is highly suspect. See reports of the state comptroller's office in *Appendix to Senate and House Journals, Tennessee, 1859–1860* (Nashville: E.G. Eastman & Co., 1860), pp. 26–31; *Appendix to Senate Journal, Tennessee, 1867–1868* (Nashville: n.p., 1868), pp. 55–7.

22. Regression analysis of household-level data for 1860 reveals a high correlation between value of personal estate and number of slaves owned, especially, as one would expect, in Middle and West Tennessee. Variations in slaveownership explain only 46 percent of variations in personal estate in East Tennessee but 77 percent in West Tennessee and fully 87 percent in Middle Tennessee.

the wealth gap between the Middle and West Tennessee counties had vanished completely, while the disparity with East Tennessee, though still large proportionally, had nonetheless narrowed considerably. In absolute terms, the wealth advantage of the typical western farmer over his eastern counterpart had fallen by 85 percent.

War, Emancipation, and the Distribution of Wealth, 1860–1870

That there was a substantial interregional redistribution of wealth reveals nothing about how (or whether) *local* wealth distribution was altered by war and emancipation. Undoubtedly, the substantial decline in wealth that characterized all sections of the state diminished the absolute difference between rich and poor in most if not all communities. As the analysis of the patterns of antebellum wealth in Chapter 2 made apparent, however, the relationship between the level and the distribution of wealth is tenuous, indeed. Certainly, there are no *a priori* reasons to tie the extent of relative inequality or stratification to the absolute level of wealth. To assess changes in local wealthholding patterns, Table 3.2 presents weighted averages by region of county-level distributions of real, personal, and total wealth as recorded in the federal manuscript censuses of 1860 and 1870.[23]

Despite the extensive economic losses inflicted by war and emancipation, in none of the Tennessee regions did the white population experience a radical redistribution of wealth. In both East and Middle Tennessee the concentration of total wealth among white farm households declined slightly, and in West Tennessee property ownership became somewhat more concentrated. Despite losses due to war and emancipation, the elite of that section – the top 5 percent of white farm households – actually increased by nearly one-fourth their share of the real and personal property of their communities. Differences aside, in all three sections the concentration of wealth continued to be pronounced and approximated patterns characteristic of the Deep South during the antebellum period. This finding not only confirms the conclusions of several recent studies of the Black Belt but also implies that the general stability of wealthholding patterns

---

23. The statistics presented in Table 3.2 are based on an analysis of every farm household in each of the sample counties. For the definition of farm households, see Appendix A.

Table 3.2 Distribution of property among white farm households,
sample Tennessee counties, 1860-1870

	East	Middle	West
Total Wealth			
% share of top 5%:			
1860	40.7	39.6	38.9
1870	36.3	38.1	47.8
% share of bottom half:			
1860	4.9	6.4	3.8
1870	6.0	7.8	3.1
Gini coefficient:			
1860	0.69	0.67	0.70
1870	0.67	0.65	0.74
Real Estate			
% share of top 5%:			
1860	39.1	40.4	45.5
1870	41.1	40.8	47.4
% share of bottom half:			
1860	1.7	5.0	3.0
1870	0.6	3.9	1.3
Gini coefficient:			
1860	0.71	0.68	0.73
1870	0.74	0.71	0.75
Personal Estate			
% share of top 5%:			
1860	48.7	41.1	37.1
1870	32.6	37.9	56.8
% share of bottom half:			
1860	6.7	5.6	3.0
1870	12.3	12.2	4.8
Gini coefficient:			
1860	0.70	0.69	0.71
1870	0.57	0.59	0.74

Source: Eight sample counties. See text.

that characterized the Cotton South obtained in small-farm areas as well.[24] Even so, an examination of the distributions of real and personal estate separately discloses interesting distinctions among the state's regions that the statistics for total wealth do not reveal.

The Civil War had a negligible effect on landholding patterns in Tennessee. In all three regions of the state, the ownership of land actually became more concentrated during the 1860s, yet the changes were slight and statistically of no consequence. As before the war, the concentration of real estate remained substantial in every section. In 1870 the agricultural elite in each region continued to control between two-fifths and one-half of local farm land, while the bottom half of farm households were still virtually landless. The Gini coefficients of concentration for 1870, like those for 1860, differ but slightly across the state, and the coefficients computed for all three sections are well within the range scholars have computed for the Deep South.[25]

In comparison, local distributions of personal wealth underwent far more interesting – and surprising – changes. Unexpectedly, the concentration of personal estate lessened considerably in every region of the state except West Tennessee, the area most heavily committed to slavery during the antebellum period. The decline was greatest among the predominantly slaveless whites of East Tennessee, where the proportion of personal wealth held by the richest 5 percent fell by one-third, the share owned by the bottom half of households doubled, and the Gini coefficient of concentra-

24. For the South as a whole, Lee Soltow found that the concentration of total wealth among all adult males decreased slightly during the 1860s. See *Men and Wealth in the United States, 1850–1870* (New Haven: Yale University Press, 1975), pp. 100, 103. Studies of the Deep South that find either general stability in the distribution of wealth or minor increases in concentration include Jonathan Wiener, *Social Origins of the New South: Alabama, 1860–1885* (Baton Rouge: Louisiana State University Press, 1978); Randolph Campbell, *A Southern Community in Crisis: Harrison County, Texas, 1850–1880* (Austin: Texas State Historical Association, 1983), pp. 302–3; Lee M. Formwalt, "Antebellum Planter Persistence: Southwest Georgia – A Case Study," *Plantation Society* 1 (1981):410–29; Kenneth Greenberg, "The Civil War and the Redistribution of Land: Adams County, Mississippi, 1860–1870," *Agricultural History* 52 (1978):292–307; Crandall A. Shifflett, *Patronage and Poverty in the Tobacco South: Louisa County, Virginia, 1860–1900* (Knoxville: University of Tennessee Press, 1982), pp. 16–23; A. J. Townes, "The Effect of Emancipation on Large Landholdings, Nelson and Goochland Counties, Virginia," *Journal of Southern History* 45 (1979):403–12; and Michael Wayne, *The Reshaping of Plantation Society: The Natchez District, 1860–1880* (Baton Rouge: Louisiana State University Press, 1983), pp. 86–91.
25. Cf. Campbell, *A Southern Community in Crisis*, p. 300.

tion decreased from 0.70 to 0.57. A similar if less pronounced decline occurred also in Middle Tennessee, where just over one-third of farm households had possessed slaves but typically only a few. In West Tennessee, paradoxically, the emancipation decade witnessed a small *increase* in the concentration of personal wealth among whites. Although the share of the bottom half of households grew slightly, this trend was more than offset by a dramatic increase of nearly 20 percentage points in the proportion of wealth controlled by the top 5 percent. As a result, whereas the concentration of personal wealth had been literally indistinguishable among the regions in 1860, *after emancipation* West Tennessee *for the first time* exhibited a substantially more concentrated distribution of personal wealth.

This discovery prompts two observations. First, in its immediate, direct effect on the relative wealth position of the local elite, emancipation was far more damaging in areas where slaveownership was confined to a small number of households than in areas where it was widespread. In East Tennessee, for example, the effect of emancipation was limited exclusively to the top decile of farm households. Emancipation had to affect the elite disproportionately. At the other extreme, emancipation in West Tennessee directly affected the wealth position of the top seven deciles. In that section the loss of wealth due to emancipation was not limited to the richest wealthholders but actually affected two-thirds of the farm population.

Second, it is likely that in West Tennessee slave wealth comprised a smaller proportion of total personal wealth among the elite than among slaveholders generally. When West Tennessee physician Samuel Oldham died in 1861, his estate included ninety-two slaves but also over $4,600 in cash on hand, as well as notes and accounts due in excess of $24,000. Similarly, Richard Taylor (John Alexander Taylor's father), who owned sixty-six slaves in 1860, died during the last year of the war, leaving over $3,400 in gold as well as other personal (nonslave) property valued at over $10,000. John Houston Bills owned eighty-nine slaves in 1860 but also held outstanding notes exceeding $50,000, Tennessee state bonds in the amount of $30,000, and nearly $15,000 in railroad stocks. Like Bills, West Tennessee's largest slaveholders regularly owned cotton gins and saw or grist mills. Several invested heavily in municipal bonds. (St. Louis Water Works Bonds seem to have been a favorite among western planters.) Over one-half of Haywood's antebellum elite owned stock in the Memphis and Ohio Railroad. At least four were merchants before the war, and two owned hotels. Although slaves had been the most valuable "property" of the West Tennessee elite, their economic activities had been wide-ranging and their "portfolios" had been diverse. Emancipation seriously depleted but did not exhaust the personal wealth of men such as these. In this they were different from the vast majority of smaller slaveholders, a difference that helps to explain how the West Tennessee elite were able to increase their share of personal wealth in the aftermath of emancipation.[26]

Geographic Mobility, 1860–1880

As argued in the preceding chapter, although cross-sectional wealth data yield valuable insight concerning the socioeconomic structure of a given population, they clearly tell only part of the story, and not even the most interesting part at that. The figures presented in Table 3.2 regarding the level and distribution of wealth in 1860 and 1870 represent mere "snapshots" taken at two discrete points in time. They are silent regarding individual experiences during the war decade. Longitudinal analysis that traces individual farmers across time allows a fuller and more accurate evaluation of the war's impact on the white agricultural population.

As was true of the 1850s, the extensive turnover among white farmers during the 1860s greatly complicates an examination of economic mobility. It is difficult to imagine a set of circumstances more disruptive of population stability than those spawned by the Civil War. The years 1861–5 were marked by the near total mobilization of adult white males across the state for military service; ultimately, more than 80 percent volunteered or were conscripted into either the Union or Confederate Army. Thus the war drew more than 100,000 Tennesseans away from their homes, many of whom would not have had occasion to leave oth-

26. See inventories (among many) of Samuel Oldham, Richard Taylor, James Bond, W. A. Anthony, Nathaniel T. Perkins, Sugars McLemore, Henry Johnson, James C. Coggeshall, and Austin Mann in Inventories of Estates, mf. roll no. 25, Office of the Circuit Court Clerk, Haywood County Records, TSLA; John Houston Bills Diary, 1 January 1860; TSLA, RG 5, Internal Improvements, Railroads, series 1, subseries 1. Other studies suggesting that antebellum planters pursued diverse economic activities include Randolph Campbell, *A Southern Community in Crisis*, p. 87; Carl N. Degler, *Place Over Time: The Continuity of Southern Distinctiveness* (Baton Rouge: Louisiana State University Press, 1977), p. 57; Paul D. Escott, *Many Excellent People: Power and Privilege in North Carolina, 1850–1900* (Chapel Hill: University of North Carolina Press, 1985), p. 5; Lacy K. Ford, *Origins of Southern Radicalism: The South Carolina Upcountry, 1800–1860* (New York: Oxford University Press, 1988), pp. 64–5; Morton Rothstein, "The Antebellum South as a Dual Economy: A Tentative Hypothesis," *Agricultural History* 41 (1967):373–82; and Rothstein, "The Cotton Frontier of the Antebellum United States: A Methodological Battleground," *Agricultural History* 44 (1970): 149–65. The argument here does not rule out the possibility that wealthy farmers elsewhere in the state also engaged in diverse economic pursuits. Many doubtless did.

Table 3.3 Geographic persistence among heads of white farm
households, sample Tennessee counties, 1850-1880 (percentages)

	East	Middle	West
All farm households			
1850-1860	55.8	47.8	36.6
1860-1870	51.0	51.5	45.5
1860-1880	57.7	54.3	42.3
Landowners			
1850-1860	61.5	51.5	44.7
1860-1870	56.0	57.2	50.4
1870-1880	65.6	59.6	50.9
Landless			
1850-1860	48.5	38.0	22.0
1860-1870	43.3	38.7	31.0
1870-1880	48.6	45.3	31.3

Source: Eight sample counties. See text.

erwise. A conservative estimate would be that one-fifth of these were killed in battle or died of disease. Most of those who survived likely returned to their homes after the war, at least temporarily, although their exposure to broader worlds may have left them predisposed for subsequent migration.[27]

During the decade of the Civil War roughly one-half of white household heads in Tennessee either died or migrated from their communities, a proportion that held approximately in every section of the state (see Table 3.3).[28] Considering the severe dislocation precipitated by the war, the level

27. Cimprich, *Slavery's End in Tennessee*, p. 22; *Tennesseans in the Civil War* (Nashville: Civil War Centennial Commission, 1964), part I, p. 1; Maris A. Vinovskis, "Have Social Historians Lost the Civil War? Some Preliminary Demographic Speculations," *Journal of American History* 76 (1989):34–58; Fred A. Bailey, *Class and Tennessee's Confederate Generation* (Chapel Hill: University of North Carolina Press, 1987), pp. 124–7. Although accurate state-specific casualty figures are not available, overall roughly one-fourth of Confederate soldiers and one-sixth of Union soldiers died while in service. Undoubtedly, the persistence figures given in Table 3.3, which apply to household heads only, are affected less by war-related mortality than if they were based on all adult males. See Vinovskis, "Have Social Historians Lost the Civil War?", p. 40.

28. Statistics presented in Tables 3.3 to 3.5 are based on a persistence analysis of heads

of persistence indicated is surprisingly high. Indeed, a comparison with rates for the 1850s shows that population stability actually increased during the 1860s in two of the state's three regions. The increase was quite small in Middle Tennessee and exhibited primarily among those who owned land. It was far more pronounced in West Tennessee, where the rate of persistence rose overall by almost one-fourth and among the landless by more than 40 percent. In contrast, persistence declined noticeably in East Tennessee, the rate falling over 5 percentage points among both landless and landowning farmers. The combined effect of these changes was to narrow substantially the range of persistence patterns across the state; the difference between the highest and lowest regional rates dropped from more than 19 percentage points during the 1850s to only 6 points during the war decade.

Without the extensive qualitative evidence necessary to establish individual motives for migration, it is impossible to determine precisely what these changes imply about the war's impact on the state's rural communities. As noted in Chapter 2, social and economic factors probably affected mobility patterns in complex and conflicting ways. All other things equal, one would expect population persistence to be highest in areas where perceptions of economic opportunity were greatest, yet extreme poverty also may have promoted high levels of persistence by erecting economic barriers to migration. On the other hand, noneconomic factors, such as the pull of kinship or institutional ties, may have commonly mitigated, if not offset entirely, the influence of economic circumstances. For the decade of the 1860s, of course, the war itself complicates matters further. If persistence levels for a given locale rose or fell significantly, it is difficult to know with certainty whether the change reflected a permanent social or economic alteration in the structure of the community or merely a momentary demographic disruption resulting from the military conflict.[29]

of farm households randomly sampled from the 1850, 1860, and 1870 censuses. For a fuller discussion see Appendix A.

29. I have not undertaken the monumental task of determining whether casualty rates varied substantially among the sample Tennessee counties, for to do so would have required sifting through well over 16,000 individual service records. It is my impression that all three regions mobilized more or less equivalently. Although many unionists in East Tennessee initially desired simply to sit out the war – mountaineers at a rally in Johnson County, for example, resolved "to be passive spectators" – Confederate confiscation and conscription acts pushed most males of military age into one army or the other. Because the casualty rate was significantly lower in the Union army, East Tennessee may have suffered less war-related mortality than the other regions of the state. See *Tennesseans in the Civil War*, part I, pp. 417–440;

The latter was probably the case with regard to East Tennessee. Economic privation resulting directly from the war may have been responsible for a portion of the significant decline in persistence within the region. Not only did military service siphon off more than half of all farm workers in the area, but in contrast to Middle and West Tennessee, there was no large pool of slave laborers to take up the slack. Food supplies thus plummeted, even before the ravages of contesting armies depleted them further. As early as 1863 a Union officer reported that upper East Tennessee had been "literally eaten up." In 1865 a limited corn crop (due to the manpower shortage) and the near total failure of the wheat crop increased suffering throughout the region. Although the Freedmen's Bureau and the East Tennessee Relief Association both provided aid to the destitute, only areas easily accessible by rail received significant assistance. For example, for supplies to reach Johnson County in extreme upper East Tennessee, they first had to be shipped by train to the nearest railhead in Abingdon, Virginia, and then hauled by wagon across the mountains some twenty miles to the county seat. Starving Johnson County farmers may well have chosen migration over dependence on relief supplies that came slowly when they came at all.[30]

The bitter internal conflict prompted by the war must also explain some portion of the decline in persistence that characterized the region. When war began, East Tennessee was far more divided in sympathy than other sections of the state. Although strongly unionist overall, its population included a large and disproportionately influential secessionist minority. (East Tennesseans split 68:32 in the June 1861 referendum.)[31] The result was an internal civil war that bred enduring animosity and countless acts of personal violence and intimidation. As long as Confederate forces retained

Brownlow's Knoxville Whig, 11 May 1861; Bryan, "The Civil War in East Tennessee," pp. 75–105.

30. An official of the East Tennessee Relief Association reported in 1865 that "the only counties to which no access has been possible with supplies . . . are Carter, Johnson, and Hancock [all in upper East Tennessee]." See Hume, *Report to the East Tennessee Relief Association at Knoxville*, p. 8. See also Bryan, "The Civil War in East Tennessee," pp. 132–63; and J. E. Jacobs to Clinton B. Fisk, 28 May 1866, mf. roll no. 38, "Records of the Assistant Commissioner for the State of Tennessee." Jacobs traveled throughout East Tennessee during the spring of 1866 on a fact-finding mission for the Assistant Commissioner but reported that he did not visit Johnson County because of its remoteness from the railroad.

31. Campbell, *The Attitude of Tennesseans toward the Union*, pp. 292–4; James W. Patton, *Unionism and Reconstruction in Tennessee, 1860–1869* (Chapel Hill: University of North Carolina Press, 1934), pp. 51–2; *East Tennessee: Historical & Biographical* (Chattanooga: A. D. Smith & Co., 1893), p. 470.

control of the region, outspoken unionists were roundly abused and many were driven from their homes and their families. As a Greeneville unionist testified, they "were persecuted like wild beasts by the rebel authorities, and . . . every imaginable wrong was inflicted upon them."[32]

In turn, Confederate sympathizers were commonly forced to leave their homes – or were not permitted to return to them – once Union control was finally reestablished. Confederate veteran William Taylor, for instance, left his Hamblen County home for Kentucky shortly after the war because "the disbanded federal soldiers in this part of East Tenn. soon made it unsafe for ex Confederates to remain." The bitter feelings of unionists in Hamilton County convinced Agustus Gothard to move to Georgia. "Bush-whackers" and similar "unfavorable conditions" forced other returning re-bels to relocate in Virginia and Illinois. A New Yorker visiting the region was horrified to learn that "the discharged Tennessee Soldiers and the low class of Citizens" were attacking returning Confederates and ordering them to leave within five days, "just because they were what they term here Southern men."[33]

Even local agents for the R. G. Dun Mercantile Agency bore witness to the extreme hostility harbored against East Tennessee Confederates. En-tries such as the following appeared frequently in the company's Greene County ledgers in the fall of 1865:

> J. W. Jackson: Rebel rascal[,] had to flee from the county to save his life. [G]one to parts unknown, entirely worthless[,] property destroyed & taken by the army.
>
> Cummings, Anderson & Co.: Mean rebels, have fled from the county, entirely worthless[,] gone to parts unknown[.]

32. Testimony of D. T. Patterson in *Report of the Joint Committee on Reconstruction*, H. R. 30, 39th Cong., 1st sess. [serial 1273], 1866, pp. 115–16. On the internal civil war in East Tennessee, see Samuel Milligan Memoirs, mf. acc. no. 13, TSLA; William E. Sloan Diary, mf. acc. no. 154, TSLA; Thomas B. Alexander, "Neither Peace Nor War: Conditions in Tennessee in 1865," *East Tennessee Historical So-ciety's Publications* 21 (1949):41–9; Bryan, "The Civil War in East Tennessee," pp. 163–82; Alden B. Pearson Jr., "A Middle-Class, Border-State Family During the Civil War," in Edward Magdol and Jon L. Wakelyn, *The Southern Common People: Studies in Nineteenth–Century Social History* (Westport, CT: Greenwood Press, 1980), pp. 155–79; and Samuel C. Williams, ed., "Journal of Events [1825–1873] of David Anderson Deaderick," *East Tennessee Historical Society's Publications* 9 (1937):107.
33. Dyer and Moore, comps., *Tennessee Veterans Questionnaires*, vol. 5, pp. 2041, 2036; vol. 3, p. 932; vol. 2, p. 760; Correspondence, Andrew Johnson, Maria S. Wofford to Andrew Johnson, 11 September 1865, TSLA.

John C. Martin: Fool rebel. Run off[,] dare not ever return, now entirely broken up, sold a valuable farm for confederate notes[.]

West & Bros.: All rebels[,] forced to leave the state, now gone to Rome Georgia, are still worth some property[.]

W. H. Henderson: Rebel, entirely worthless forced to leave the county to save his life[.]

J. W. R. Doak: A worthless rebel, fled to Va, whence he dare never return[.][34]

It was probably only mild hyperbole, then, when Greene County politician David Patterson reported to the Joint Committee on Reconstruction that "we have now but few rebels in East Tennessee." Even so, animosities appear to have died down rather quickly, and the forced exile of Confederates was relatively short in duration.[35] Evidence from the 1870s reinforces the conclusion that the pattern of the war decade was not only unusual but also temporary. Persistence rates between 1870 and 1880 rose significantly among both landowning and landless farmers – the overall figure for the region closely approximated that for the 1850s – and the sharp contrast with the Black Belt counties that had characterized the antebellum years once again emerged.

In contrast, the Civil War had a more lasting demographic effect in both Middle and West Tennessee. In both regions the trend toward higher persistence initiated during the 1860s continued throughout the subsequent decade as well. In Middle Tennessee persistence was even higher during the 1870s than during the 1860s, and in West Tennessee the elevated persistence patterns of the 1860s continued basically unchanged.[36] Why persistence levels should have risen in the western two-thirds of the state is

---

34. Tennessee Vol. 12, pp. 217–72, R. G. Dun & Co. Collection, Baker Library, Harvard University Graduate School of Business Administration. For a discussion of the Dun records and of the Mercantile Agency's operations in the South, see James H. Madison, "The Credit Reports of R. G. Dun & Co. as Historical Sources," *Historical Methods Newsletter* 8 (1975):128–31; and James D. Norris, *R. G. Dun & Co., 1841–1900: The Development of Credit-Reporting in the 19th Century* (Westport, CT: Greenwood Press, 1978).

35. On this point, see William E. Sloan Diary, pp. 106–9; O. R. Broyles to T. A. R. Nelson, 15 August 1865, 22 January 1866, T. A. R. Nelson Papers, East Tennessee Historical Society; Dyer and Moore, comps., *Tennessee Veterans Questionnaires*, vol. 4, pp. 1349–50, 1464; vol. 5, pp. 2036, 2040–1, 2245; Samuel W. Scott and Samuel P. Angel, *History of the 13th Regiment Tennessee Volunteer Cavalry* (Philadelphia: P. W. Ziegler & Co., 1903), p. 417.

36. The slight drop in the overall persistence rate in West Tennessee during the 1870s stemmed from an increase in the ratio of landless to landowning households between 1860 and 1870. Persistence among the two subcategories was virtually identical from decade to decade.

subject for speculation. The most obvious possibility, though ultimately unprovable, is that perceptions of economic opportunity improved after the Civil War. Scholars have frequently hypothesized that emancipation was a boon to slaveless whites as well as to slaves, allowing those who owned land to compete more effectively for laborers and those who did not to compete more effectively for jobs.[37] Tennessee veterans interviewed in their old age disagreed as to whether the presence of slavery adversely affected white opportunity. A large majority maintained that economic chances had been good in their communities and that upward mobility had been extensive. A minority, however, complained that their opportunities were limited because "slaveholders did not have to higher mutch extry laber." If not for slavery, a poor Middle Tennessean recalled, "a man working as I was could have secured better wages."[38]

Although opinions were mixed, this much is clear: If the coexistence of slave and free labor had not been a "moral impossibility"[39] before the war, the two sources of manpower had definitely varied inversely across the state. After emancipation, however, major slaveholding areas appear to have become significantly more appealing to middle- and lower-class whites. This was particularly true of West Tennessee, where, despite the disruption of the Civil War, the persistence rate among household heads who had not owned slaves when the war began (both landowners and landless) was nearly 60 percent higher during the 1860s than for the slaveless during the preceding decade.[40]

Evidence concerning immigration, an element historians rarely explore, further indicates that both Middle and West Tennessee attracted immigrating whites far more after emancipation than before. In both regions the white farm population in the sample counties had declined absolutely during the 1850s (by approximately 10 percent in Middle Tennessee and 12 percent in West Tennessee), yet mushroomed dramatically after 1860. Over the next twenty years the number of white farm households grew by 42

---

37. See, for example, John S. Ezell, *The South Since 1865* (New York: Macmillan Publishing Co., 1975), pp. 222–3; Roger Ransom and Richard Sutch, *One Kind of Freedom: The Economic Consequences of Emancipation* (Cambridge: Cambridge University Press, 1977), p. 104; Fred A. Shannon, *The Farmer's Last Frontier: Agriculture, 1860–1897* (New York: Rinehart & Winston, 1945), p. 80.
38. Dyer and Moore, comps., *Tennessee Veterans Questionnaires*, vol. 4, p. 1709; vol. 2, p. 807.
39. John E. Cairnes, *The Slave Power: Its Character, Career, and Probable Designs* (London: Macmillan & Company, 1863), 2d rev. ed., pp. 147, 143.
40. Persistence increased from 22.1 percent during the 1850s to 35.2 percent during the 1860s.

percent in the major cotton-producing counties and by an astounding 64 percent in the Central Basin counties. Obviously, part of the growth was a product of natural increase and part also reflected the increased persistence among farmers in both regions, yet a large part also resulted from a substantial influx of white households into the counties. Compared with the 1850s, the ratio of in-migrating household heads to household heads disappearing from the county (due to death or migration) doubled during the 1860s in Middle Tennessee's Wilson County and tripled in West Tennessee's Haywood. In the latter county the number of newcomers by 1870 equaled two-thirds of the entire 1860 white population![41]

Absolute Economic Mobility and the Structure of the White Farm Population, 1860–1880

Although admittedly circumstantial, such changing patterns of geographic mobility constitute persuasive evidence that white impressions regarding the extent of economic opportunity in heavily black areas shifted substantially after emancipation. This fact is crucial in and of itself, given the axiom that "the definition of southern reality . . . begin[s] with the southern perception of reality."[42] Even so, it is important to know the degree to which such altered expectations were realized.

To answer this question it is again useful to apply the distinction between absolute and relative economic mobility employed in the evaluation of the 1850s. To review, absolute mobility occurred solely whenever a persisting household head crossed from either direction the line between landowners and landless. (The acquisition or loss of slaves need no longer be considered.) Relative mobility, in contrast, took place when persisters experienced a change in wealth level relative to their neighbors but without necessarily crossing obvious economic boundaries. Although the former constituted a clear-cut change in socioeconomic status, the status significance of the latter is problematic and likely varied from case to case.

White migrants attracted to Middle and West Tennessee probably believed that the possibility of acquiring a farm – the likelihood of upward absolute mobility, in other words – had increased since before the war. Veterans interviewed long after the war rarely compared the antebellum

41. These figures were determined by tracing sample 1870 household heads to the 1860 census.
42. J. Mills Thornton III, *Politics and Power in a Slave Society: Alabama, 1800–1860* (Baton Rouge: Louisiana State University Press, 1977), p. 479.

Table 3.4 Upward mobility into the landowning ranks among heads of white farm households, sample Tennessee counties, 1850-1880 (persisters only)

	East	Middle	West
Percentage of 1850 landless owning land in 1860	44.4	53.8	65.1
Percentage of 1860 landless owning land in 1870	35.7	49.5	60.5
Percentage of 1870 landless owning land in 1880	43.6	47.1	50.9

Source: Eight sample counties. See text.

and postbellum periods in this regard, but a few who did maintained that the economic chances of laboring whites improved after the war. Middle Tennessean J. L. Walton, whose father had been an overseer on a large plantation, believed that "verry few" poor whites had been able to buy farms of their own "up to the time . . . the war broke out[,] but a great manny did after the war." William Dillihay, son of a small slaveholder from Maury County, similarly noted that "slaveholders before the war seem[ed] to be the leading men of the country, causing little demand for white labor. [T]he negro emancipation placed white people on a more equal footing causing much better opportunities for young men to buy homes since the war than before."[43]

Strictly speaking, Walton and Dillihay were wrong, at least with regard to the immediate postwar years (see Table 3.4). Compared with the previous decade, during the 1860s the probability that a persistent, landless farmer would acquire land declined noticeably, though not drastically, in every region of the state. This said, it was still true that land acquisition among persisting farmers was proportionally much more likely in Middle than in East Tennessee, and higher still in the major cotton counties of West Tennessee. To the degree that this was true, of course, the state simply mimicked patterns that had also characterized the 1850s.

Although the relative order among the regions remained the same during the 1870s as well, the large gap that had separated them since the 1850s narrowed substantially. The rate of land acquisition in East Tennessee in-

43. Dyer and Moore, comps., *Tennessee Veterans Questionnaires*, vol. 5, p. 2133; vol. 2, p. 692.

creased by nearly 8 percentage points during the 1870s, returning to what might be considered the antebellum norm for that section. Elsewhere in the state, however, the likelihood of farmownership continued to fall; the decline was minor in Middle Tennessee but quite severe in the counties farther west. During the previous two decades, high cotton prices had translated into high incomes for West Tennessee farmers, including those who rented farms from others while saving to purchase their own. Cotton prices had skyrocketed during the war, exceeding $1 per pound by 1864. Even though they had begun to fall steadily once peace had been restored, as late as 1870 West Tennessee farmers could still sell their cotton in Memphis for twenty-four cents a pound, more than double its antebellum price. Cotton prices continued their steep descent during the 1870s, however. Indeed, they actually fell briefly below antebellum levels, dropping by more than 50 percent between 1872 and 1878. Although the high prices of the 1850s and 1860s must have made it easier for the landless to acquire farms of their own, the far lower prices of the late 1870s must have made farm ownership more difficult.[44]

At the same time that it was becoming more challenging for the landless to acquire farms, it was also becoming harder for those who owned farms to retain them (see Table 3.5). The proportion of landowning farmers losing all their real estate increased during the 1860s in every section, rising most noticeably in East Tennessee, where it grew by two-thirds, and in West Tennessee, where it approximately doubled. Over the course of the following decade, however, the incidence of such downward mobility dropped significantly in these two regions while continuing to rise in Middle Tennessee. The result was that in East and West Tennessee during the 1870s the frequency with which farmers lost their farms was only marginally higher than it had been before the Civil War, whereas in Middle Tennessee such instances were nearly twice as common.[45]

44. For the 1870 price in Memphis, see Memphis *Public Ledger*, 5 January 1870. When weighted by the Warren-Pearson Wholesale Price Index to control for inflation, the 1870 price still represented an advance in real value of 64 percent over that for 1859. See U.S. Bureau of the Census, *Historical Statistics of the United States, Colonial Times to 1957* (Washington, DC: Government Printing Office, 1960), p. 115. For cotton prices throughout the postbellum period, see Matthew B. Hammond, *The Cotton Industry: An Essay in American Economic History* (New York: The Macmillan Company, 1897).

45. There are few comparable studies to provide a benchmark for comparison, but see statistics for Clarke County, Georgia, in Frank Jackson Huffman Jr., "Old South, New South: Continuity and Change in a Georgia County, 1850–1880" (Ph.D. diss., Yale University, 1974), pp. 107–13.

Table 3.5 Downward mobility into the landless ranks among heads of white farm households, sample Tennessee counties, 1850-1880 (persisters only)

	East	Middle	West
Percentage of 1850 landowners owning no land in 1860	6.6	6.3	7.8
Percentage of 1860 landowners owning no land in 1870	10.8	8.1	15.1
Percentage of 1870 landowners owning no land in 1880	8.1	11.9	10.8

Source: Eight sample counties. See text.

Evaluated as a whole, the two decades after 1860 were years not only of decreased opportunity among tenants and farm laborers but also of greater insecurity among farmowners. If anything, changes in absolute economic mobility during the 1860s were greatest in the predominantly white, small-farming region of East Tennessee, certain testimony to the bitter internal civil war that was visited on the region. The disruption engendered there seems to have been relatively short-lived, however. By the 1870s, patterns of both geographic and absolute economic mobility closely approximated those for the last antebellum decade. In the other two sections of the state, on the other hand, the trends initiated by war and emancipation were more enduring. Fully fifteen years after the war's conclusion, landownership in both Middle and West Tennessee remained a more formidable goal than during the halcyon days of the 1850s, and the danger of losing one's land continued to be more threatening.

All other things equal, the combined effect of these trends should have been to increase the proportion of landless households within the farm populations of Middle and West Tennessee. Because the federal census stopped asking respondents for information on property ownership after 1870, precise estimates of landlessness among the *entire* farm population are quite difficult to construct for subsequent years. It is actually easier, on the other hand, to estimate landlessness among farm *operators* only because the census in 1880 began to record tenure information for all individuals enumerated on the agricultural schedules. Efforts to construct estimates for the entire white farm population are complicated enormously

Table 3.6 Percentage distribution of heads of white farm households, with lower- and upper-bound tenancy rates, sample Tennessee counties, 1880

	East	Middle	West
Appearing on both schedules I and III			
(1) Owner-operators	61.2	61.6	65.6
(2) Definite tenants	17.7	20.0	31.7
Appearing on schedule I only			
(3) Farmers and farm laborers	21.2	18.4	2.7
Estimated tenancy rate (percentage)			
Lower bound[a]	22.4 (+5.7)	24.5 (+3.4)	32.6 (+18.9)
Upper bound[b]	38.8 (-0.9)	38.4 (+9.3)	34.4 (+9.2)
Mobility-adjusted upper bound[c]	36.3 (-3.4)	35.2 (+6.1)	29.9 (+4.7)

[a]Equivalent to $[(2)/(1)+(2)]$; "level II" as defined by Bode and Ginter.   [b]Equivalent to $\{[(2)+(3)]/[(1)+(2)+(3)]\}$; "level IV" as defined by Bode and Ginter.   [c]Projected upper-bound tenancy rate had absolute economic mobility patterns of the 1850s obtained through 1880.
Source: Eight-county sample, see text.
Note: Figures in parentheses represent percentage-point change since 1860.

by the presence of black farmers on the agricultural schedules, which did not indicate the race of enumerated operators, as well as by the continued numerical importance of farmers and farm laborers listed on the population schedules only. By sifting through the population census to determine the number of white and black farm households, and by matching a sample drawn from the agricultural census to the population census to determine the racial and occupational breakdown of farm operators, it is possible to arrive at fairly precise race-specific figures regarding the structure of the entire farm population (see Table 3.6). Unfortunately, this indirect method of estimating the farm population not listed on the agricultural schedules provides no means of ascertaining the proportion of such farmers and farm laborers who were landless. In order not to minimize the increase in land-lessness following the war, the tenancy rates given in Table 3.6 are based on the assumption that *all* such households were landless, thus exaggerating (probably slightly) the true proportion of farm households without land. (For a full explanation of the estimating process, see Appendix A.)

To facilitate comparison with 1860, Table 3.6 presents both lower- and upper-bound estimates of the rate of tenancy for each region in 1880. The lower-bound figures reflect the proportion of farm operators listed in the agricultural schedules – now numbered as schedule III – who owned no land. They are comparable to the published "tenancy rates" commonly analyzed by scholars of postbellum agriculture. As explained in Chapter 1, however, because some portion of the farmers and farm laborers omitted from schedule III may actually have been tenants or sharecroppers, these lower-bound figures likely underestimate the extent of landlessness among farm operators. The upper-bound estimates, on the other hand, take into account not only the operators enumerated on schedule III but also all farm household heads listed only on schedule I, some proportion of whom were unenumerated operators. Although the latter estimates undoubtedly exaggerate considerably the true rate of tenancy among operators, they should closely approximate the overall proportion of landlessness in each farm population. Because of inconsistency among census enumerators in distinguishing between farm operators and farm laborers, the latter figures are preferable for comparisons among regions and between census years.

As anticipated, the estimates in Table 3.6 reveal that the proportion of farm households without land rose significantly after 1860 in both Middle and West Tennessee. Had the analysis been confined to the farm operators listed on the agricultural census, however (as postbellum studies typically are), it would have appeared that the shock inflicted by war and emancipation was far more severe in West Tennessee than elsewhere. The rate of tenancy among enumerated farm operators more than doubled in the western cotton counties but rose only modestly in the other regions. When the entire farm population is taken into account, however, the impression that emerges is much different. The overall proportion of landless households actually declined slightly in East Tennessee and increased by approximately the same proportion, one-third, in both of the other sections. Close examination shows that in West Tennessee a large part of the drastic increase after 1860 in the lower-bound rate of tenancy reflected a relative upward shift among the landless from wage labor to tenancy, rather than a downward shift from farmownership.[46]

46. Thus confirming the speculation of Roger Ransom and Richard Sutch. See *One Kind of Freedom*, p. 104. In West Tennessee the proportion of landless farm house-

Nor did the increase in overall landlessness that characterized both Middle and West Tennessee result primarily from the alteration in patterns of absolute economic mobility after 1860 (i.e., the greater difficulty farmers faced in both acquiring and holding on to land). In addition to the upper- and lower-bound tenancy rates, Table 3.6 also presents mobility-adjusted rates, the tenancy rates projected for each region if the incidence of absolute economic mobility (upward and downward) had remained constant between 1850 and 1880 at 1850s levels. By controlling for changes in economic mobility, the adjusted estimates allow an evaluation of the importance of population growth – due to natural increase, higher geographic persistence, and increasing in-migration – as a factor underlying the growing landlessness of the postwar period.[47] These figures suggest that the proportion of landless households would have increased substantially in both Middle and West Tennessee between 1860 and 1880 (by 6 and 5 percentage points respectively), even if the incidence of upward mobility had not declined nor the frequency of downward mobility increased. Historians have frequently misinterpreted the increase in tenancy rates that characterized every southern state after Appomattox, assuming that it reflected the widespread dispossession (or "proletarianization") of smallholding farmers. In the postbellum South, according to this common view, tenancy, "which might have marked the first step up the agricultural ladder[,] signaled instead the loss of land by former owners."[48] This was far from the case in Middle and West Tennessee, however, where a booming increase in the white farm population after 1860 severely strained the capacity of both regions to accommodate young aspiring farmowners, and explains between one-half and two-thirds of the increase in the rate of landlessness.[49] Quite

hold heads appearing on the agricultural schedule increased from approximately 43 percent in 1860 to 92 percent twenty years later.

47. Declining mortality was not likely a factor, at least not during the war decade when it undoubtedly rose above normal levels.

48. Harold Woodman, "Post-War Southern Agriculture and the Law," *Agricultural History* 53 (1979):319–37 (quotation on p. 337). Steven Hahn employs the term "proletarianization" in *The Roots of Southern Poverty: Yeoman Farmers and the Transformation of the Georgia Upcountry, 1850–1890* (New York: Oxford University Press, 1983), p. 162. See also Barbara Jeanne Fields, "The Nineteenth-Century American South: History and Theory," *Plantation Society in the Americas* 2 (1983): 7–27; Roark, *Masters Without Slaves*, p. 193; and Forrest McDonald and Grady McWhiney, "The South from Self-Sufficiency to Peonage: An Interpretation," *American Historical Review* 85 (1980):1111. Significantly, none of these authors has undertaken the sort of longitudinal analysis necessary to validate the hypotheses that they present as fact.

49. Cf. Gavin Wright, *Old South, New South: Revolutions in the Southern Economy Since the Civil War* (New York: Basic Books, Inc., 1986), p. 112.

literally, the growing landlessness of both sections was less an indication of economic malaise than of the degree to which they had become economically more attractive.

Relative Economic Mobility during the 1860s

Despite undeniable alterations in the incidence of absolute mobility, it should be clear that war and emancipation did not lead to a revolution in ownership within any of Tennessee's major divisions. Still, there may have been a more subtle reordering of farmers that the foregoing analysis would not detect. The examination of absolute mobility, after all, measures only the most obvious instances of socioeconomic change and describes the experiences of but a small minority of persisting household heads. The vast majority of persisters did not move between landownership and landlessness during the period covered. It is also desirable, then, to gauge the extent of relative economic mobility in each region among the local farm populations as a whole. Imitating the method employed in Chapter 2, each persisting farmowner has been accorded a decile ranking for real wealth for both 1860 and 1870 based on the value of his property in relation to the overall distribution of real estate among landowners within the local county. Farmers who were landless in either or both years received decile ranks of zero for the appropriate year. Taking advantage of the additional wealth information available after 1850, decile rankings for total wealth (real + personal) have also been assigned to *each* farm household head. (Thus there is no rank of zero for total wealth.) Unfortunately, because of the absence of satisfactory wealth data after 1870, it is necessary to limit the analysis to the 1860s, the decade most directly affected by war and emancipation.

Despite the unprecedented physical destruction and social disruption inflicted by the American Civil War, an astonishing degree of stability characterized the wealthholding continuum in every section of the state. This was particularly true with regard to landholding patterns (see Table 3.7). Farmers who owned little or no land before the war were likely to move upward somewhat, but otherwise exceptional stability was the rule. This does not indicate stagnation among persisters during the 1860s but, rather, the absence of clear trends. Indeed, typically two-thirds or more experienced at least some movement – the overall mean was around two deciles per persister – yet there were no strong tendencies in such movements. There were no trends – whether toward overall improvement, deterioration, or leveling – in any of the state's regions. Relative mobility was

Table 3.7 Movement of persisting white farmers in **real** wealth position relative to local white farm population, sample Tennessee counties, 1860-1870 (heads of household only)

	East	Middle	West
Real Wealth decile in 1860	Median real wealth decile in 1870		
0	0	0	1
1	2	2	2
2	3	2	2
3	5	3	6
4	5	5	3
5	6	5	5
6	6	7	5
7	7	7	7
8	8	8	9
9	9	9	9
10	10	10	10
Mean absolute decile change	1.8	1.8	2.2
Percentage moving 0-1 deciles	61	57	50

Source: Eight-county sample, see text.
Note: Farmers with "0" real wealth decile rank were landless. Farmers with "1" rank were in the poorest decile of landowners; farmers with rank of "10" were in the wealthiest.

more or less randomly distributed and typically far slighter than the mean figures on decile change would suggest; in every section of the state between 50 and 61 percent of persisting household heads moved only one decile or none at all.

Unexpectedly, this impression of stubborn continuity is scarcely diminished when changes in personal wealth (which had included slaves) are incorporated into the evaluation. Table 3.8 shows changes in the total (real + personal) wealth ranking of persisting farmers during the 1860s. As was the case with real wealth, farmers who began the decade with little or no total wealth had moved upward substantially by 1870. (This reflected more than any other factor the frequent acquisition of land among those ranked in the lower deciles. The acquisition of even a tiny farm was typically sufficient to move the successful new owner above the bottom one-third of wealthholders.) There was no corresponding downward trend within the

Table 3.8 Movement of persisting white farmers in
total wealth position relative to local white farm population, sample Tennessee counties, 1860-1870
(heads of household only)

	East	Middle	West
Total Wealth decile in 1860	*Median total wealth decile in 1870*		
1	3	3	4
2	3	3	4
3	4	4	5
4	5	5	6
5	6	6	7
6	7	8	7
7	8	8	8
8	8	9	9
9	9	9	10
10	10	10	10
Mean absolute decile change	1.5	1.7	2.1
Percentage moving 0-1 deciles	65	56	50

Source: Eight-county sample, see text.
Note: Farmers with "1" rank were in the poorest decile of household heads; farmers with rank of "10" were in the wealthiest.

top half of the white farm population, however. Despite their extensive
losses during the war decade, in every region farmers situated in the upper
deciles maintained their positions with an astounding perseverance.

This latter finding is worth expanding on in some detail. Historical research since the 1970s has shown that large planters in the Cotton South
held onto their lands with great tenacity, but little if anything is known
about the experience of local elites outside the Black Belt. In fact, evidence
from the sample counties suggests that in Tennessee at least the ability of
local elites to survive war and emancipation with their relative position
intact varied negligibly between plantation and nonplantation regions. Considering persisters only, the proportion of the top 10 percent of 1860
wealthholders (in value of total wealth) who remained in their lofty position
a decade later varied by but a few percentage points, ranging from 65
percent in East Tennessee to 68 and 73 percent in Middle and West Tennessee, respectively. Although their level of wealth had been vastly reduced,

and many were haunted by debts to a degree unknown in antebellum years, stable, persistent agricultural elites continued to characterize all regions of the state during the decade of the Civil War.

Conclusion

In the final analysis, the American Civil War did alter wealthholding patterns across rural Tennessee, but in complex ways that defy easy generalizations regarding continuity and change. By erasing nearly $100 million in human property, emancipation initiated a sweeping *interregional* redistribution of wealth among the state's white agricultural population and largely eliminated the pronounced disparities in overall wealth level that had distinguished the three grand divisions up to 1860. At the local level, however, the drastic reductions in *levels* of wealth did not translate into significantly more egalitarian *distributions* of wealth. The distribution of real estate among whites was almost literally unaffected during the war decade, whereas the concentration of personal estate lessened noticeably in every region except the one most heavily dependent on slavery. As a result, the concentration of total wealth among the white farm population declined slightly in East and Middle Tennessee, while increasing slightly in the Black Belt counties of West Tennessee. Such minor changes notwithstanding, one conclusion is abundantly clear: From one end of the state to the other, inequality proved to be an amazingly durable phenomenon that withstood the cataclysm of civil war without diminution.

As before the war, however, pronounced inequality continued to be consistent with widespread, modest upward economic mobility in all regions of the state. The disruption of internal civil war seems to have affected economic mobility patterns only temporarily in Appalachian East Tennessee, but in the remainder of the state landless farmers found it consistently more difficult to acquire farms and landowning farmers found their hold on the land more tenuous. Despite these adverse tendencies, however, the former plantation regions actually became more attractive to migrating whites during the period; the proportional increase in landlessness that characterized both Middle and West Tennessee actually reflected their growing appeal to already landless farmers rather than the proletarianization of smallholders.

The cumulative effect of these changes, significantly, was to lessen even further the minor variations in wealthholding patterns that had distinguished the state's three major regions during the late antebellum period.

Prior to the American Civil War they had differed substantially in reliance on slavery, in commercial orientation, and in levels of wealth and income; during the war they had varied significantly in extent of physical destruction and loss of human property. After the war's conclusion, however, in the local concentration and distribution of wealth and the pace at which it was acquired or forfeited, Tennessee's white farm communities were virtually indistinguishable from the Appalachians to the Mississippi River.

4. "Change and Uncertainty May Be Anticipated": Freedmen and the Reorganization of Tennessee Agriculture

In an 1883 report to the Tennessee legislature, state commissioner of agriculture Joseph Buckner Killebrew addressed widespread complaints concerning the unreliability of black labor. "Our labor system," he explained, "as it regards the farm, may be regarded as passing through a transition period, and gradually approaching its permanent condition. Until this is reached, change and uncertainty may be anticipated."[1] Killebrew made this observation two decades after the Emancipation Proclamation, and, if correct, the commissioner's assessment indicates that "change and uncertainty" were persistent rather than transient features of the postbellum Tennessee countryside, enduring for a generation after the slaves were freed.

Killebrew's conclusion is troubling, for it clashes with basic tenets of the conventional wisdom among scholars in regard to the postbellum transformation of southern agriculture. During the past twenty years, many economists and historians have produced works that examine the social and economic reconstruction of the South and the fate of the freedmen in

1. *Biennial Report of the Commissioner of Agriculture for the State of Tennessee, 1881–1882* (Nashville, 1883), p. 9. A modified version of this chapter has appeared previously as "Freedmen and the Soil in the Upper South: The Reorganization of Tennessee Agriculture, 1865–1880," *Journal of Southern History* 59 (1993):63–84. That article, reflecting work at a much earlier stage of the larger research project, was based on nine sample counties rather than eight. Problems with antebellum data sources later prompted me to drop two West Tennessee counties (Gibson and Hardeman) and substitute Fayette in their place. The estimates presented in this chapter for West Tennessee thus differ (albeit slightly) from those in the journal article.

particular.[2] The account emerging from these studies posits a fleeting transitional period (several years at most) after emancipation – during which new institutions and patterns of behavior were quickly but firmly established – followed by seventy-five years with little change. Agreement is so widespread in regard to basic factual details of the account that a standard scenario is accepted by the majority of scholars. Its fundamental elements are well known and need be reviewed here only briefly. Immediately after emancipation white landowners sought to impose upon the freedmen a system of gang labor in which black laborers were compensated with either a fixed cash wage or a share of the crop.[3] This arrangement, however, quickly proved unsatisfactory to landowners and freedmen alike. White employers became disillusioned with the system after disastrous crop years in 1866 and 1867. The freedmen were openly hostile from the beginning to a system that so closely resembled slavery. Given the extensive demand for labor during the early postbellum years, the freedmen were able to force their disenchanted employers to experiment with a variety of other arrangements.

According to this standard scenario, the period of experimentation led rapidly to the predominance of sharecropping – commonly defined as a labor arrangement in which *"individual family units*, in payment for their labor on a separate parcel of land, receive[d] a share of the output produced on that parcel of land."[4] Scholars debate whether the transition from gang labor to sharecropping was immediate and direct or whether it occurred in stages, but few deny that sharecropping emerged decisively as "the South's

2. For recent surveys of the literature see Harold D. Woodman's "Economic Reconstruction and the Rise of the New South, 1865–1900," in John B. Boles and Evelyn Thomas Nolen, eds., *Interpreting Southern History: Historiographical Essays in Honor of Sanford W. Higginbotham* (Baton Rouge and London: Louisiana State University Press, 1987), pp. 254–307; and Lee J. Alston, "Issues in Postbellum Southern Agriculture," in Lou Ferleger, ed., *Agriculture and National Development: Views on the Nineteenth Century* (Ames: Iowa State University Press, 1990), pp. 207–28.

3. Until recently the standard scenario held that fixed-wage labor wholly predominated in the immediate aftermath of emancipation, a contention that both Ralph Shlomowitz and Gerald David Jaynes have disproven. See Shlomowitz, "The Transition from Slave to Freedman: Labor Arrangements in Southern Agriculture, 1865–1870" (Ph.D. diss., University of Chicago, 1979), pp. 39–55; Jaynes, *Branches Without Roots: Genesis of the Black Working Class in the American South, 1862–1882* (New York: Oxford University Press, 1986), p. 45.

4. Ralph Shlomowitz, "The Origins of Southern Sharecropping," *Agricultural History* 53 (1979):557–75 (quotation on p. 575, emphasis added).

new peculiar institution."[5] The decline of the plantation system and the rise of family-based tenancy were "both swift and thorough," and the process of institutional reorganization was essentially complete by 1880, if not much earlier.[6] Throughout this economic transformation, the freedmen's lack of capital and credit, combined with the hostility of whites to black landownership, severely restricted their acquisition of property. Indeed, two prominent historians of the period maintain that "the option of owning agricultural land was one that may be disregarded when considering the situation faced by the freedmen in the Reconstruction Period."[7]

5. Harold D. Woodman, "Post–Civil War Southern Agriculture and the Law," *Agricultural History* 53 (1979):319–37 (quotation on p. 326).
6. Roger L. Ransom and Richard Sutch, *One Kind of Freedom: The Economic Consequences of Emancipation* (Cambridge: Cambridge University Press, 1977), pp. 12, 56 (quotation).
7. Ibid., p. 87. Other works advancing the standard scenario, in whole or in part, include Vernon Lane Wharton, *The Negro in Mississippi, 1865–1890* (Chapel Hill: University of North Carolina Press, 1947), pp. 62–71; George Brown Tindall, *South Carolina Negroes, 1877–1900* (Columbia: University of South Carolina Press, 1952), pp. 92–100; Robert Cruden, *The Negro in Reconstruction* (Englewood Cliffs, NJ: Prentice-Hall, Inc., 1969), pp. 40–4; Lawrence D. Rice, *The Negro in Texas, 1874–1900* (Baton Rouge: Louisiana State University Press, 1971), pp. 151–69; Peter Kolchin, *First Freedom: The Responses of Alabama's Blacks to Emancipation and Reconstruction* (Westport, CT: Greenwood Press, 1972), pp. 30–48; James Roark: *Masters Without Slaves: Southern Planters in the Civil War and Reconstruction* (New York: W. W. Norton & Co., 1977), pp. 142, 158; Jay R. Mandle, *The Roots of Black Poverty: The Southern Plantation Economy After the Civil War* (Durham, NC: Duke University Press, 1978), pp. 17–19; Jonathan M. Wiener, *Social Origins of the New South: Alabama, 1860–1885* (Baton Rouge: Louisiana State University Press, 1978), pp. 66–73; Gilbert Fite, *Cotton Fields No More: Southern Agriculture, 1865–1980* (Lexington: The University Press of Kentucky, 1984), pp. 2–5; Gavin Wright, *Old South, New South: Revolutions in the Southern Economy Since the Civil War* (New York: Basic Books, Inc., 1986), pp. 84–7; Stephen V. Ash, *Middle Tennessee Society Transformed, 1860–1870: War and Peace in the Upper South* (Baton Rouge: Louisiana State University Press, 1988), pp. 186–8; Eric Foner, *Reconstruction: America's Unfinished Revolution, 1863–1877* (New York: Harper & Row, 1988), pp. 403–9; and Roger L. Ransom, *Conflict and Compromise: The Political Economy of Slavery, Emancipation, and the American Civil War* (Cambridge: Cambridge University Press, 1989), pp. 235–40. A few scholars have directly challenged one or more of the fundamental elements of the accepted wisdom. See Shlomowitz, "Transition from Slave to Freedman" and "Origins of Southern Sharecropping"; Lee J. Alston and Robert Higgs, "Contractual Mix in Southern Agriculture Since the Civil War: Facts, Hypotheses, and Tests," *Journal of Economic History* 42 (1982):327–53; Nancy Virts, "Estimating the Importance of the Plantation System to Southern Agriculture in 1880," *Journal of Economic History*, 47 (1987):984–8. Notwithstanding these challenges, there is a near universal presentation of the standard scenario in even the most recent textbook surveys of United States his-

This chapter reevaluates the standard scenario and its implications for the economic history of African Americans by examining the postbellum reorganization of agriculture in Tennessee during the first fifteen years after emancipation. It focuses on two elements of the conventional account: the rapidity of institutional reorganization resulting in the clear predominance by 1880 of sharecropping among freedmen, and the negligible extent of land accumulation by blacks during the same period. In contrast to the standard scenario, a thorough longitudinal study of Tennessee farmers brings new light to bear on the important questions of black mobility, acquisition of agricultural land, and persistence of landholding.

Competing Visions in the Aftermath of Emancipation

As in many sections of the South, freedom came gradually to Tennessee's 275,000 slaves. Invading Union armies occupied much of Middle and West Tennessee before the war was scarcely a year old and began actively to recruit local slaves as soldiers and laborers by the spring of 1863. Slaves in the remote reaches of upper East Tennessee, on the other hand, had to wait until February or March of 1865 for the advent of Union forces.[8] Variations in timing aside, by the spring of 1865 emancipation had profoundly disrupted long-established labor patterns in all three sections of the state. "All the traditions and habits of both races had been suddenly

tory. See, e.g., John M. Blum et al., *The National Experience: A History of the United States,* 7th ed. (New York: Harcourt Brace Jovanovich, 1989), pp. 358–9; Joseph R. Conlin, *The American Past: A Survey of American History,* 3rd ed. (New York: Harcourt Brace Jovanovich, 1990), pp. 565–6; Robert A. Divine et al., *America, Past and Present,* brief 2d ed. (Glenview, IL: Scott, Foresman and Company, 1990), pp. 274–5; James A. Henretta, *America's History* (Chicago: The Dorsey Press, 1987), pp. 500–2; Jonathan Hughes, *American Economic History,* 3d ed. (New York: Harper Collins, 1990), pp. 257–9; Winthrop D. Jordan and Leon F. Litwack, *The United States* (Englewood Cliffs, NJ: Prentice-Hall, 1991), pp. 409–11; Gary B. Nash et al., *The American People: Creating a Nation and a Society* (New York: Harper & Row, 1986), pp. 546–8; Mary Beth Norton et al., *A People and a Nation: A History of the United States,* 3d ed. (Boston: Houghton Mifflin Company, 1990), pp. 457, 488–9. See also William J. Cooper Jr. and Thomas E. Terrill, *The American South: A History* (New York: McGraw-Hill, 1991), pp. 423–5.

8. Ash, *Middle Tennessee Society Transformed,* pp. 83–5; Joseph H. Parks, "A Confederate Trade Center under Federal Occupation: Memphis, 1862–1863," *Journal of Southern History* 7 (1941):289–314; Charles Faulkner Bryan, "The Civil War in East Tennessee: A Social, Political, and Economic Study" (Ph.D. diss., University of Tennessee, 1978); John Cimprich, "Slavery's End in East Tennessee," *East Tennessee Historical Society's Publications* 52–53 (1980–1981):78–88.

overthrown," a former slaveholder from Middle Tennessee recalled, "and neither knew just what to do, or how to accommodate themselves to the new situation."[9] Both freedmen and former masters responded in numerous ways to this uncertainty, but in the course of their accommodation three primary, competing visions of the postwar world emerged.

Tennessee's freedmen, for their part, were unequivocal in their desire to own land and homes of their own and envisioned a black community of freeholding farmers largely independent of white interference. Clinton B. Fisk, assistant commissioner of the Freedmen's Bureau for the state of Tennessee, encapsulated this vision nicely in a letter to a northern acquaintance in the summer of 1865:

> Twenty, thirty or forty acres of land, a mule or two, plows, seeds and hoes, a cabin with his own family therein, a schoolhouse for him and his nearby, and freedom, is the picture the late slaves are painting for the future. God grant that they may realize all this at no distant day, and let all the people say amen.[10]

Initially the freedmen expected the federal government to facilitate this dream through the redistribution of their masters' plantations. "This was no slight error, no trifling idea," a Freedmen's Bureau official explained, "but a fixed and earnest conviction as strong as any belief a man can ever have." Although forced ultimately to relinquish the hope of federal intervention, they nonetheless held tightly throughout the reconstruction era to the vision of an independent black yeomanry. Tennessee delegates to a regional meeting in 1871 decried the "spirit of land monopoly" in the state that closed "avenues of prosperity and personal independence." Similarly, a black convention meeting in Nashville at the end of the decade resolved that "the first want of the colored laborer . . . is to become a land holder." A group of local blacks affirmed this pronouncement, asserting that "the very foundations of the social, intellectual and material advancement of a race depends upon their becoming possessors of the soil."[11]

9. James Buckner Killebrew, "Recollections of My Life," p. 160, James Buckner Killebrew Papers, Southern Historical Collection, typescript copy in Tennessee State Library and Archives.

10. Clinton B. Fisk to Lyman Abbott, 11 August 1865, quoted in *New York Times*, 14 August 1865, p. 5.

11. Leon Litwack, *Been in the Storm So Long: The Aftermath of Slavery* (New York: Vintage Books, 1979), p. 399; *Proceedings of the Southern States Convention of Colored Men* (Columbia, SC: Carolina Printing Company, 1871), p. 25; Nashville *Daily American*, 9 May, 17 August 1879. See also testimony of Brevet Major General Edward Hatch, in United States Congress, *Report of the Joint Committee on*

Former slaves were essentially unified in their vision, but white reactions fell into one of two broad categories. Many white Tennesseans wanted nothing to do with free black labor and looked toward the day in which blacks would play no role in the state's agriculture. Future commissioner of agriculture James B. Killebrew, for example, expressed relief at the passing of the "peculiar institution" but denigrated the potential of free black workers. Describing his slaves as a "constant responsibility . . . [that] was well nigh intolerable," Killebrew assured his children that "I never was free until my slaves were free." The freedmen could not be trusted, however, because "no reliance can be placed in the disposition of the negroes to fulfill their contracts." Killebrew spoke for many ex-masters in calling for a new system of agriculture founded on "the skill and labor of intelligent white men."[12]

Strategies for the realization of this vision were typically two-fold, calling for the colonization of the freedmen in conjunction with the immigration of white laborers to the state. Although newspapers across the state called for the forced removal of ex-slaves, this theme resounded most loudly in predominantly white East Tennessee. The Greeneville New Era, for example, confessed to support "the freedom of all slaves . . . in whose nostrils is the breath of life," but, it continued,

> We are [also] for the colonization of the Freedmen, after we have one grand, universal jubilee, in some country of their own, without mixing, mingling or commingling, believing as we most religiously do, that such a course of policy will be for the best interest of both the white and the black man.[13]

The state's governor, William G. Brownlow, seconded such proposals, urging that the state "provid[e] for them [the freedmen] a separate and appropriate amount of territory, and settl[e] them down permanently as a nation of freedmen."[14]

Reconstruction, House Report Number 30, 39th Cong., 1st sess. (Washington, DC: Government Printing Office, 1866), part I, p. 107; and of Moses Singleton, recounted in Nell Irvin Painter, Exodusters: Black Migration to Kansas after Reconstruction (New York: W. W. Norton & Co., 1976), pp. 113–17.

12. Killebrew, "Recollections of My Life," pp. 160, 340.

13. Greeneville New Era, 12 August 1865. For other examples see ibid., 2, 23 September 1865; Brownlow's Knoxville Whig and Rebel Ventilator, 23 August 1865; Whitelaw Reid, After the War: A Tour of the Southern States, 1865–1866 (New York: Moore, Wilstach & Baldwin, 1866; reprint ed., New York: Harper & Row, 1965), p. 352.

14. Quoted in William Gillespie McBride, "Blacks and the Race Issue in Tennessee Politics, 1865–1876" (Ph.D. diss., Vanderbilt University, 1989), p. 39.

Those who would eschew black labor entirely recognized the necessity of developing a pool of white laborers sufficient to replace the former slaves. Except in East Tennessee, where landless whites were numerous, proponents of white labor sought a solution from outside the South. Ex-Confederate general Gideon Pillow, for instance, recruited northern whites for one of his numerous plantations and confessed himself "anxious to try the system of white labor of that character." Other Tennesseans appealed specifically to the "industrious Germanic race" to replace "the now indolent negro." The Tennessee legislature formalized such efforts shortly after the war, creating a Board of Immigration to advertise opportunities for northern and foreign laborers and to invite "every industrious immigrant to take up his abode in this State."[15]

A majority of white landowners, though, dismissed the twin goals of black colonization and white immigration as impractical and unnecessary and expected, or at least hoped, to continue their reliance on black labor. Many who had owned slaves were gripped with a sense of helplessness at the loss of their "servants" and went to great lengths – including deception, cajolery, and violence – to retain their services. Robert Falls's former owner, an East Tennessean named Goforth, held several of his younger slaves for almost a year after the war's end by convincing them that they were bound to him until age twenty-one.[16] Adopting a different strategy, Middle Tennessean Paulding H. Anderson, who had owned thirty-two slaves in Wilson County in 1860, wrote to his former slave Jourdon in the summer of 1865, requesting that he return to his old home and promising him good wages and fair treatment.[17] At the other extreme, Hardeman County planter Amos Black relied on tactics of intimidation, shooting one of his former slaves and assuring the rest that "there is no law against

15. *Bolivar Bulletin*, 8 December 1866; Greeneville *New Era*, 21 October 1865; Gideon Pillow to O. O. Howard, 22 December 1865, quoted in *New York Times*, 6 January 1866; *New Yorker Staats Zeitung*, quoted in *New York Times*, 8 October 1865; Hermann Bokum, *The Tennessee Handbook and Immigrant's Guide* (Philadelphia: J. B. Lippincott & Co., 1868), p. 141. See also "Word to Immigrants" in Killebrew, *Introduction to the Resources of Tennessee*, pp. 385–404. Similar calls echoed across the southern states. For examples, see *Southern Cultivator* 25 (1867), no. 1, p. 1; no. 3, p. 69; no. 8, p. 247.

16. Norman R. Yetman, ed., *Life Under the "Peculiar Institution": Selections from the Slave Narrative Collection* (New York: Holt, Rinehart and Winston, Inc., 1970), p. 118.

17. Paulding Anderson's letter does not survive, but see the former slave's reply, Jourdan Anderson to P. H. Anderson, preserved in Lydia Maria Child, *The Freedmen's Book* (Boston: Ticknor and Fields, 1865; reprint ed., New York: Arno Press, 1968), pp. 265–7.

killing niggers and I will kill every d----d one I have if they do not obey me and work just as they did before the war."[18]

Despite the diversity of reaction among former slaveholders, most ex-masters entered the postwar world with the firm conviction that freedmen would not work effectively without coercion and that, in order to secure a reliable black work force, it would be necessary to restrict black mobility and to fashion land and labor arrangements that resembled slavery as closely as possible. West Tennessee planter John Houston Bills, for example, considered the freedmen a "lazy insolent race" who "wanted freedom only to loaf and do nothing. . . . Not one in a dozen [would] make a living without the lash or a certainty of it if they do not work." Echoing these sentiments, James A. Rogers, a former Haywood County slaveholder, wrote to the state's governor demanding legal restrictions on black migration. It would "never do," he argued, "to suffer them to roam about at will, go where they please, work or let it alone as they please. A negro must have someone to manage him, and he must be required to respect and obey his employer."[19]

Ultimately, of course, none of these competing visions was fully realized. The freedmen's dream of an independent black yeomanry was essentially stillborn. It was formally laid to rest by President Johnson's amnesty proclamation of May 1865, which promised full restoration of property, except slaves, to all who would take an oath of loyalty to the union.[20] Realistically, however, it had already been effectively undermined by the Republican Party's free labor ideology, which combined a commitment to property rights with the conviction that indiscriminate charity would "demoralize" any freedman willing to work.

In Tennessee the task of practically applying such principles was allotted

18. Affidavit of Joe Black, quoted in John Cimprich, *Slavery's End in Tennessee, 1861–1865* (University: University of Alabama Press, 1985), p. 120.

19. John Houston Bills Diary, Southern Historical Collection (microfilm copy in Tennessee State Library and Archives), entry for 30 August 1864; James A. Rogers to William G. Brownlow, 15 May 1865, quoted in James W. Patton, *Unionism and Reconstruction in Tennessee* (Chapel Hill: University of North Carolina Press, 1931), p. 153. Cf. *The Fayetteville Observer*, 16 November 1865; Greeneville *New Era*, 9 September 1865; *Bolivar Bulletin*, 5 January 1867; *New York Times*, 9 July 1865, 6 January 1866.

20. Technically, Johnson's proclamation excepted several categories of prominent Confederates, but the vast majority of those excluded – approximately 13,500 out of about 15,000 applicants – ultimately received individual pardons from the president. See James M. McPherson, *Ordeal by Fire: The Civil War and Reconstruction*, 2d. ed. (New York: McGraw-Hill, Inc., 1992), pp. 496, 502.

to Clinton B. Fisk, the ranking Freedmen's Bureau official in the state. A New York native with long-standing abolitionist sympathies, Fisk was undoubtedly sincere in his support of the freedmen. Indeed, he frequently addressed northern audiences concerning the plight of the former slaves and enjoyed describing himself as "one who has marched with them through the Red Sea of strife, sympathized with them in all their sufferings, [and] labored incessantly for their well-being."[21] His genuine concern notwithstanding, Fisk spent much of his tenure with the Freedmen's Bureau shutting down "contraband" camps, encouraging blacks to remain with their former masters, and convincing them – with difficulty – that the federal government had no intention of giving them land.[22] Speaking to audiences of freedmen across the state, the assistant commissioner espoused the virtues of "economy and well-directed industry" and assured the freedmen that "there is no obstacle which persistent work will not remove out of your way." Convinced that they would work industriously if treated honestly by their employers, Fisk resolved to secure for the freedmen a fair chance – and nothing more. As he explained to a Brooklyn audience, further assistance was unnecessary: "Give [the freedmen] fair play and the whole industrial problem is solved."[23]

If fair play alone was less than the freedmen had hoped for, white landowners who had envisioned an exclusively white labor force were likewise disappointed. The large-scale colonization of freedmen proved to be financially, logistically, and politically unfeasible, and the anticipated influx of white immigrant labor, Germanic or otherwise, failed to materialize. (The appeals to immigrants, which amounted at bottom to a plea to "come take the place of our former slaves," could not have been terribly enticing.) Finally, former masters seeking the immobilization of black labor and the reimposition of slavery under another name were similarly unsuccessful,

---

21. Clinton B. Fisk, *Plain Counsels for Freedmen in Sixteen Brief Lectures* (Boston: American Tract Society, 1866; reprint ed., New York: AMS Press, Inc., 1980), p. 5; Alphonso A. Hopkins, *The Life of Clinton Bowen Fisk* (New York: Funk & Wagnalls, 1888; reprint ed., New York: Negro Universities Press, 1969), pp. 28–33, 94–107; Paul David Phillips, "A History of the Freedmen's Bureau in Tennessee" (Ph.D. diss., Vanderbilt University, 1964), p. 47.

22. *New York Times*, 2 September, 22 November 1865; Fisk, *Plain Counsels for Freedmen*, pp. 12, 20.

23. Fisk, *Plain Counsels for Freedmen*, pp. 60, 20; *New York Times*, 22 November 1865. See also Fisk's "Circular # 2," quoted in Phillips, "A History of the Freedmen's Bureau in Tennessee," p. 117. On northern free-labor ideology, see Eric Foner, "Reconstruction and the Crisis of Free Labor," in Foner, *Politics and Ideology in the Age of the Civil War* (New York: Oxford University Press, 1980), pp. 100–2.

deterred primarily by the free labor commitment of Republicans in Congress. If fair play did not mean land redistribution, neither did it mean immobilization and effective serfdom.[24]

Black Geographic Mobility

This final point bears repeating. It is crucial to recognize that, throughout the Reconstruction era, landowners failed miserably in their efforts to limit black movement. Although historians universally acknowledge extensive black mobility during the first months after emancipation, some scholars maintain that immobility characterized black agricultural laborers after the first few years of freedom. Despite emancipation, they argue, freedmen remained tied to one place, restricted by the coercive tactics of landowners and merchants.[25] Careful statistical analysis, however, demonstrates that landless blacks were highly mobile in every section of the state and, in East and Middle Tennessee, actually more mobile than whites in similar economic circumstances. For at least the first fifteen years after emancipation, evidently, Tennessee blacks moved freely throughout the countryside, exploiting their newfound mobility to reunite with long-lost relatives, to take advantage of economic opportunities, and to resist exploitation.

Although Freedmen's Bureau agents repeatedly stressed that "you can be as free and as happy in your old home . . . as anywhere else in the world," few of the former slaves accepted their argument. As throughout the South, emancipation in Tennessee set off a migratory wave as freedmen deserted their former masters in countless personal declarations of inde-

24. In contrast, opposition by state Republicans was halfhearted at best. In the summer of 1865 a stringent Black Code very nearly passed the Republican-dominated Tennessee General Assembly. On state politics during the Reconstruction period, see McBride, "Blacks and the Race Issue in Tennessee Politics"; Patton, *Unionism and Reconstruction in Tennessee*; Thomas B. Alexander, *Political Reconstruction in Tennessee* (Nashville: Vanderbilt University Press, 1950); and Theodore B. Wilson, *The Black Codes of the South* (University: University of Alabama Press, 1965).

25. See, for example, Crandall Shifflett, *Patronage and Poverty in the Tobacco South: Louisa County, Virginia, 1860–1900* (Knoxville: University of Tennessee Press, 1982), p. 25; Mandle, *The Roots of Black Poverty*, p. v; Roark, *Masters Without Slaves*, pp. 142–3; Wiener, *Social Origins of the New South*, pp. 69–70; William Cohen, "Negro Involuntary Servitude in the South, 1865–1940," *Journal of Southern History* 42 (1976):31–60. Cohen modifies his position in a recent monograph, the fullest treatment of black geographical mobility to date. See *At Freedom's Edge: Black Mobility and the Southern White Quest for Racial Control, 1861–1915* (Baton Rouge: Louisiana State University Press, 1991).

pendence. Recalling the first few months of freedom, an East Tennessee freedman remembered "how the roads was full of folks walking and walking along when the niggers were freed. Didn't know where they was going. Just going to see about something else somewhere else."[26] For many freedmen this meant nothing more than the migration to a neighboring farm or plantation, but thousands traveled long distances in search of loved ones, and thousands more sought the economic opportunities and military protection afforded by cities and larger towns.

This rural–urban shift occurred in each region of the state but was strongest in Middle and West Tennessee. To cite the most extreme case, the black population of Memphis increased from around 3,900 in 1860 to more than 16,000 only five years later. The black population of Nashville grew less spectacularly but still increased by 150 percent during the decade of the 1860s. A rapid influx of blacks also characterized the smaller towns of Gallatin to the north and Columbia to the south. In early July 1865 Middle Tennessee planter Nimrod Porter noted that all of his former slaves had stopped work and gone to Columbia. A week and a half later Porter traveled there himself. His diary for July 15 contains the single terse entry: "I went to town. Many negroes." By 1870 blacks constituted 44 percent of the population of all Middle Tennessee towns.[27]

Although they failed to maintain such a frenetic pace, Tennessee blacks continued to be highly mobile throughout the 1870s (see Table 4.1). Specifically, of black farmers and farm laborers sampled from eight rural counties in 1870, only about two-fifths persisted (i.e., remained) in their counties

26. Yetman, *Life Under the "Peculiar Institution,"* p. 118.
27. Cimprich, "Slavery's End in East Tennessee"; Bobby L. Lovett, "Memphis Riots: White Reactions to Blacks in Memphis, May 1865–July 1866," *Tennessee Historical Quarterly* 38 (1979):10; Ash, *Middle Tennessee Society Transformed*, pp. 183–4; Fred S. Palmer to H. S. Bowen, 30 June 1866, and Ben P. Runkle to Clinton B. Fisk, 4 April 1866 in "Records for the Assistant Commissioner for the State of Tennessee," Record Group 105, Bureau of Refugees, Freedmen, and Abandoned Lands, NARA, microcopy T142; Nimrod Porter Diary, Southern Historical Collection (microfilm copy in Tennessee State Library and Archives). Freedmen flocked to the cities not only to await the distribution of land but also for the very good reason that there were limited job opportunities in the countryside in 1865. During the last springtime of the war, most Tennessee farmers had planted only partial crops, reducing the demand for labor, and those who did require laborers generally had no means with which to pay them. See "Will the Freedmen Work?", *Report of the Commissioner of Agriculture for the Year 1865* (Washington, DC: Government Printing Office, 1866), pp. 135–6; Phillips, "A History of the Freedmen's Bureau in Tennessee," p. 128; *New York Times*, 2 September 1865.

Table 4.1 Geographic persistence among heads of farm households, sample Tennessee counties, 1870-1880 (percentages)

	East	Middle	West
Whites	57.7	54.3	42.3
Landowners	65.6	59.6	50.9
Landless	48.6	45.3	31.3
Blacks	43.4	41.6	38.8
Landowners	48.0	47.1	54.2
Landless	43.0	41.0	38.6

Source: Eight-county sample, see text.

of origin ten years later, a pattern that obtained consistently across the state.[28] This estimate exaggerates the extent of *inter*county mobility; some of the absent freedmen had surely died by 1880 and others were simply missed by the census enumerator or undetected by the matching process. However, because the persistence figures necessarily ignore instances of *intra*county mobility – such farm-to-farm movement was likely far more common than longer migrations[29] – they still undoubtedly constitute lower-bound estimates of black movement during the decade.

A comparison with the persistence figures for white farmers shows plainly that, overall, freedmen were noticeably *more* mobile in all three regions of the state. The appropriate implication to be drawn from this is less certain, however. Without question, the common scholarly descriptions of black laborers as bound to the land or "in a position tantamount to debt peonage" do not apply to Tennessee's freedmen during the 1870s.[30] Because historians have commonly interpreted the absence of mobility as evidence of labor exploitation, the high rates of migration among the state's

28. These figures are based on a sample of 6,042 black and white farm household heads drawn from the 1870 population census and traced to the population census for 1880. For a fuller description of the sample and method of analysis, see Appendix A.
29. Ronald L. F. Davis, *Good and Faithful Labor: From Slavery to Sharecropping in the Natchez District, 1860–1890* (Westport, CT: Greenwood Press, 1982), p. 179.
30. See Wiener, *Social Origins of the New South*, p. 70; Hahn, *The Roots of Southern Populism: Yeomen Farmers and the Transformation of the Georgia Upcountry, 1850–1890* (New York: Oxford University Press, 1983), p. 186. Hahn's description pertains to both races.

blacks might be interpreted positively. On the other hand, as Gavin Wright points out, contemporaries – unlike modern-day historians – associated mobile laborers of either color "with tramps and vagabonds, people with no ambition."[31] Furthermore, as the analysis of geographic mobility among rural whites both before and after the Civil War reveals, geographic mobility tended to be lowest among those with the greatest stake in society. With this in mind, the relatively greater restlessness among the freedmen seems less encouraging, for it likely reflected at least partially their less successful efforts at property accumulation. In East and Middle Tennessee, for instance, much of the apparently large gap between the races is erased by controlling for differences in the extent of landownership between the two groups. Evidently, if black farmers were less geographically stable than white farmers, to a large extent it was because they were less likely to own land. The same was also true in West Tennessee, where the freedmen overall were slightly less persistent than whites but were actually more persistent within each landownership category.

Toward a "New Peculiar Institution"?

For good or for ill, Tennessee's freedmen exhibited a high rate of geographic mobility from their emancipation through the end of the 1870s. As a consequence, the new land and labor arrangements then evolving were being fashioned by white landowners in need of labor and black laborers without land – but with the conspicuous capability of moving about and contracting with employers of their own choosing. This freedom to pick up stakes and leave, perhaps more than any other factor, helps explain the considerable success of Tennessee's freedmen in resisting efforts by their former masters to impose a gang-labor system on them. Finding it difficult to attract sufficient labor, numerous landowning whites acceded to black demands for increased independence by subdividing their lands into small tenant plots to be farmed by black families as tenants or sharecroppers.[32]

Some whites accepted this new approach as inevitable and adopted it

31. Wright, *Old South, New South*, p. 98.
32. For evidence of a significant labor shortage, especially severe in Middle and West Tennessee, see United States Congress, *Report of the Joint Committee on Reconstruction*, part I, pp. 100, 112, 121; in "Records of the Assistant Commissioner" see Fred S. Palmer to H. S. Bowen, 14 May 1866; Charles F. Johnson to H. S. Bowen, 8 June 1866, both on mf. roll 38; R. Caldwell to Clinton B. Fisk, 4 April 1866; J. H. Gregory to J. R. Lewis, 10 July 1866; Fred S. Palmer to S.W. Groesbeck, 11 March 1867, all three on mf. roll 40.

willingly. For example, former Confederate general Richard Ewell, who owned a large farm in Middle Tennessee, conceded to his son- in-law in early 1867 that "if one wishes to farm profitably to any extent it must be by grazing [which required a minimal labor force] or else by planting on shares." Planning for the coming year, Ewell resolved to "let out the cotton land to darkies on shares" and cultivate only thirty to forty acres of the best lands himself.[33]

Most whites who adopted the system probably did so with far greater hesitation. West Tennessee planter John Houston Bills was probably more representative. In 1866 the sixty-five-year-old Bills reluctantly consented to make a contract with his former slaves, providing each family with land, work stock, and feed. To the end he remained skeptical. "I will try this one year," he confided to his diary, "and if it prove a success may try it again, but at my time of life I have little idea of following free negroes."[34]

According to the standard scenario for the Cotton South, Ewell and Bills were *typical* in their acquiescence; by 1880, if not much sooner, the great majority of former masters, whether graciously or grudgingly, had adopted a labor system in which black families cultivated particular plots as sharecroppers or tenants rather than working for cash or share wages under central direction. Scholars commonly ground this assumption on two types of evidence. First, they cite numerous contemporary references to the expansion of tenancy and sharecropping; second, they interpret the dramatic postbellum growth in the number of farm units and the sharp simultaneous decline in farm size as indirect proof of a massive shift from wage labor to sharecropping or tenancy.[35]

Similar evidence suggests that comparable changes were occurring in Tennessee. For example, as early as 1869 a Nashville newspaper published a survey of the conditions of Tennessee freedmen and concluded that "the great bulk of colored men farm on shares." In 1874 the state's commissioner of agriculture alluded to the adoption of the "share system" and concluded that it "exists to a great extent at present." The United States Department of Agriculture corroborated this report two years later, estimating that two-thirds of Tennessee farms were "occupied on the share system."[36] Profound alterations in the number and size of farms across the

---

33. R. S. Ewell to Major Campbell Brown, 10 March 1867, file #13, box #1, Brown–Ewell Papers, Tennessee State Library and Archives.
34. John Houston Bills Diary, 30 August, 13 December 1866.
35. See, for example, Ransom and Sutch, *One Kind of Freedom*, pp. 68–71.
36. Nashville *Daily Press and Times*, 7 January 1869; Killebrew, *Introduction to the Resources of Tennessee*, p. 351; United States Department of Agriculture, *Report of*

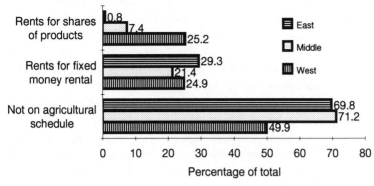

Figure 4.1 Profile of landless heads of black farm households, sample Tennessee counties, 1880

state also seem to provide additional verification. Between 1860 and 1880 the total number of farm units across the state doubled while the average size of a farm decreased from 251 to 125 acres.[37]

A Cross-section of the Black Farm Population
Yet, such evidence does not establish conclusively the predominance by 1880 of family-based labor arrangements among the state's freedmen. What is lacking is a cross-sectional analysis of the *entire* black farm population – not confined to black farm operators enumerated on the census of agriculture – that demonstrates convincingly the prevalence of tenancy, sharecropping, or both. Historians have been slow to undertake such analyses, despite their obvious value, presumably because of the considerable time and effort required to identify individuals reporting farm occupations but not listed as farm operators on the agricultural schedules.

In fact, a close examination of the 1880 black farm population in Tennessee raises considerable doubt about the supremacy of family-based labor arrangements among the state's freedmen. Figure 4.1 presents a profile of landless black farm households in the eight sample Tennessee counties.[38]

the Commissioner . . . for the Year 1876 (Washington, DC: Government Printing Office, 1877), p. 131.
37. U.S. Census Office, Tenth Census [1880], *Statistics of Agriculture* (Washington, DC: Government Printing Office, 1883), p. 25. Considering the sample counties only, the number of farms increased by 105 percent between 1860 and 1880, whereas average farm size declined by 51 percent.
38. The estimated profile is based on a second random sample of 4,761 farm operators drawn from the agricultural census for 1880 and matched to the population census

The first two groups of households consist of individuals listed on the agricultural schedule under one of the two landless headings defined (albeit ambiguously) by the census: "rents for fixed money rental" or "rents for shares of products." The third is a catchall category that includes all farm household heads (whether farmers or farm laborers) listed on the population schedule but missing on the agricultural census.

The exact composition of the second and third categories is difficult, if not impossible, to establish. Those listed as renting "for shares of products," it is safe to assume, certainly included share tenants, that is, renters who provided all the essentials for farm operation except the land itself and paid rent in the form of a share of the crops they produced. Whether this category also included sharecroppers – individuals who farmed specific plots but supplied only their labor, receiving work stock, tools, and seed from their landlord – is more problematic. Because sharecroppers resembled in fundamental respects both farm operators and farm laborers, it is possible that census enumerators excluded them from the ranks of farm operators (omitting them from the agricultural schedule) and recorded their names on the population schedule only.

There are several reasons to doubt that they did so, however, and to believe that the classification "rents for shares of products" included both share tenants and sharecroppers. First of all, the instructions to census enumerators in 1880 regarding agricultural occupations were brief and excessively vague. They consisted of the single sentence: "Be very particular to distinguish between farmers and farm laborers."[39] Intuitively, it makes sense that in attempting to follow this terse directive the easiest distinction for enumerators to have made was between individuals responsible for specific plots of land and those who were not. This would have led the census enumerators to group sharecroppers with share tenants on the agricultural schedule and to exclude only farm laborers who hired out by the day, month, or year to neighboring farm operators. It seems far less likely, on the other hand, that enumerators – who frequently had their hands full simply spelling the names of their respondents – would have consistently tried to differentiate between sharecroppers and share tenants, a process that required specific information concerning the capital and managerial inputs of respondents.[40]

for the same year. For a description of the sample and of the method of determining the farmless farm population, see Appendix A.

39. Carroll D. Wright, with William C. Hunt, *The History and Growth of the United States Census* (Washington, DC: Government Printing Office, 1900), p. 172.

40. Two of the earliest scholarly studies of postbellum agriculture both maintain that census officials typically grouped sharecroppers with share tenants. See Enoch

Also, in contrast to other southern states, in Tennessee by 1880 there seem to have been no legally recognized distinctions between share tenants and sharecroppers. In a series of decisions handed down during the 1870s the Tennessee Supreme Court defined sharecroppers as "tenants in common of the crops" and made clear that even before harvest and disposal the cropper's portion represented personal property, not wages.[41] In addition to the element of convenience, therefore, legal realities would have also prompted enumerators to include sharecroppers on the agricultural schedule along with share renters. For these reasons, then, the estimates reflected in Figure 4.1 suggest that in every section of the state *between one-half and three-fourths* of landless blacks engaged in agriculture had not yet achieved the status of tenant or sharecropper by 1880. Considering the eight sample counties as a whole, nearly three-fifths (57 percent) of landless blacks were omitted by the census of agriculture. Such freedmen either constituted an enormous body of farm operators inexplicably omitted from the agricultural schedule or, more likely, were ex-slaves who toiled for wages on farms operated by others.[42]

Understandably, one hesitates to accept the latter conclusion in light of the considerable conflicting evidence previously cited: the extensive contemporary testimony and the pronounced change in the number and size of farms. Careful scrutiny, however, shows both types of evidence to be illusory. Most of the evidence scholars cite to establish the rapid preemi-

Marvin Banks, *The Economics of Land Tenure in Georgia* (New York: Columbia University Press, 1905), p. 83; Robert Preston Brooks, *The Agrarian Revolution in Georgia* (Madison: University of Wisconsin Press, 1914), pp. 55, 79.

41. Donald L. Winters, "Postbellum Reorganization of Southern Agriculture: The Economics of Sharecropping in Tennessee," *Agricultural History* 62 (1988):1–19. The practical implications of the court's decisions may not have been great, however. Winters notes that in response to the rulings, landlords quickly began to demand that sharecroppers grant liens on their portions of the crop as security against supplies advanced. For the incorrect assertion that legal distinctions between share tenants and sharecroppers characterized every southern state, see Harold Woodman, "Post–Civil War Southern Agriculture and the Law," *Agricultural History* 53 (1979):319–37.

42. Other scholars have drawn the same conclusion about these black farmers and farm laborers without farms. See, for example, James R. Irwin, "Farmers and Laborers: A Note on Black Occupations in the Postbellum South," *Agricultural History* 64 (1990):53–60; Robert C. Kinzer, *Kinship and Neighborhood in a Southern Community: Orange County, North Carolina, 1849–1881* (Knoxville: University of Tennessee Press, 1987), p. 112; Shifflett, *Patronage and Poverty in the Tobacco South*, p. 39. Frederick Bode and Donald Ginter speculate that in 1860 Georgia "farm laborers" not included in the agricultural census may have actually been sharecroppers or tenants, but I have found no evidence to support such a conclusion. See *Farm Tenancy and the Census in Antebellum Georgia* (Athens: University of Georgia Press, 1986), especially pp. 97–100.

nence of sharecropping comes from the detailed reports of local Freedmen's Bureau agents, who both supervised and documented the transition to free labor during the first three to four years after the war's end. Reviewing such reports, the chief official of the Freedmen's Bureau in Tennessee estimated in 1866 that one-half of the state's blacks were working for a "share of the crop," yet it is almost certain that this ambiguous category included the separate systems of share wages and sharecropping, and possibly even share tenancy as well. In Wilson County, for instance, the local agent reported in the same year that two-thirds of local freedmen were working on shares, yet a detailed analysis of contracts (which the agent had personally approved) shows that only about one-fifth of former slaves in the county labored in family units.[43]

Significantly, a careful study of Bureau reports also reveals a definite trend *away* from share arrangements after the first two to three years of experimentation. According to the standard scenario, planters shifted toward sharecropping from wages because of the demands of the freedmen combined with the effects of a series of severe crop failures in 1866 and 1867, which left white landowners more than willing to share risk with their black workers. In Tennessee, however, the short harvests of the early postwar years, coupled with the frequent fraudulent division of crops under the various share systems, seem to have prompted a shift toward fixed wages in keeping with the wishes of the freedmen. This was the opinion of a Freedmen's Bureau agent stationed in Murfreesboro, who reported that numerous freedmen in Middle Tennessee who had been "cropping for the last two years now prefer to work for wages, as they can understand this mode better, and it does not give the employer the same opportunity to take advantage, as in the more complicated crop system." From all across the state during the early months of 1868 came reports that freedmen were shifting to fixed wages from "*the old plan* of a portion of the crop."[44]

43. "Report of the Assistant Commissioner for Kentucky and Tennessee," Senate Doc. 6, 39th Cong., 2d sess., p. 130; Fred S. Poston to H. S. Brown, 23 June 1866, mf. roll 38, "Records of the Assistant Commissioner"; entry 3513, "Register of Contracts for Wilson County," Record Group 105, Bureau of Refugees, Freedmen and Abandoned Lands, NARA.

44. Fred S. Palmer to S. W. Groesbeck, 10 February 1868 (quotation, emphasis added); Palmer to W. H. Bowen, 15 April 1868, mf. roll 41, "Records of the Assistant Commissioner"; United States Congress, "Freedmen's Affairs in Kentucky and Tennessee," House Ex. Doc. 329, 40th Cong., 2d sess. (Washington, DC: Government Printing Office, 1868), pp. 28–31. For evidence of a similar shift at approximately the same time in South Carolina, see John Scott Strickland, "Traditional Culture and Moral Economy: Social and Economic Change in the South

Unfortunately, because the Bureau had almost completely phased out its operations in Tennessee by 1869, it is impossible to determine either the extent or the duration of this shift toward fixed wages. Persistent contemporary references to "shares" during the late 1860s and 1870s suggest that the trend may have been short-lived, yet they do not establish conclusively that sharecropping became the alternative of choice. Most such references, as Ralph Shlomowitz has observed with regard to the South generally, "do not unambiguously refer to sharecropping contracts." Rather, they may instead indicate the renewed vitality of collective share-wage arrangements in which squads, or small gangs, labored for their employer in return for a share of the harvest.[45] Consider, for example, the 1869 survey revealing that the "great bulk" of Tennessee freedmen farmed "on shares": The same report also explained that under this system the ex-slaves were frequently organized into "clubs" that received half of all they produced. Similarly, the 1876 USDA report that testified to the prevalence of "the share system" in Tennessee went on to criticize the "wastefulness" of that system, "especially with *large gangs of hands*."[46] At best, then, early postbellum references to the share system or farming on shares constitute highly ambiguous and unconvincing proof of the widespread adoption of sharecropping across the state by 1880.

More difficult to dismiss are the undeniable changes in the number and size of farm units across the state. A careful examination of population trends, however, reveals two frequently ignored factors that help to reconcile such aggregate changes with the continued vitality of wage labor among the freedmen. The first trend, and the most important for the state as a whole, involved the population of white farm operators, which continued to grow – dramatically so – during the two decades after 1860. As shown in the preceding chapter, this was partly a function of population growth, but in East and West Tennessee at least it also reflected a relative shift among landless whites from agricultural wage labor to some form of tenancy. As a result, although the total number of farm operators in the

Carolina Low Country, 1865–1910," in Steven Hahn and Jonathan Prude, eds., *The Countryside in the Age of Capitalist Transformation: Essays in the Social History of Rural America* (Chapel Hill: University of North Carolina Press, 1985), p. 157.

45. Shlomowitz, "The Transition from Slave to Freedman," pp. 128–9.
46. Nashville *Daily Press and Times*, 7 January 1869; United States Department of Agriculture, *Report of the Commissioner . . . for the Year 1876*, p. 138, emphasis added. The state Commissioner of Agriculture's remarks concerning the "share system" are similarly ambiguous and may be interpreted as a reference to the system of share wages, not sharecropping. See Killebrew, *Introduction to the Resources of Tennessee*, p. 351.

Table 4.2 Increase in farm operators, sample Tennessee counties, 1860-1880

	East	Middle	West
Increase in **total** operators	2,363	3,434	4,352
Increase in **white** operators	2,176	2,401	1,216
Increase in white operators as proportion of total increase	92%	70%	28%

Source: Eight-county sample, see text.

sample counties increased by 105 percent between 1860 and 1880, the number of white operators alone grew by 57 percent, thus accounting for more than half of the total change (see Table 4.2).

A second crucial trend obtained among the state's black population, which not only underwent rapid growth but also exhibited a substantial spatial redistribution as the center of gravity of the black population shifted noticeably westward. The proportion of the state's black population living in West Tennessee increased from approximately 36 percent in 1860 to 43 percent twenty years later.[47] The importance of this trend is demonstrated by developments in the heavily black sample counties of the western region. Between 1860 and 1880 the number of farms in Fayette and Haywood counties increased by a factor of nearly 3.7. Because the ratio of the slave to free farm population just before emancipation was approximately 1.6:1, a reasonable conclusion would be that an overwhelming majority of the former slaves had become farm operators. Although sensible, the inference would be erroneous for two reasons. To begin with, the number of *white*-operated farms had increased by 74 percent. If every freedman in the region had remained a wage laborer, the number of farm units still would have nearly doubled.[48] The growth of white-operated units was proportionally far less important in the western sample counties than elsewhere, however, explaining just 28 percent of the area's spectacular increase in farms. More

47. Derived from United States Census Office, Tenth Census [1880], vol. I, *Statistics of Population* (Washington, DC: Government Printing Office, 1883), pp. 407–8. For evidence of a similar trend in Alabama, see Kolchin, *First Freedom*, pp. 12–19.

48. The number of white-operated farms increased by 72 percent in the sample East Tennessee counties and by 48 percent in the sample Middle Tennessee counties.

important was the 73 percent growth in the counties' black population. To accommodate every western black farm household in 1880, the number of farm units needed to have been still 50 percent greater, even after the fantastic increase of the previous twenty years. Put another way, the number of farms would have had to increase by a factor of 5.4 after 1860. Although changes in the number and size of southern farms may provide an accurate index of the decline of the plantation system, it is evident that they are an incomplete and highly inaccurate indicator of the growth of black tenancy and sharecropping.

Black Land Accumulation

In sum, then, the standard scenario for the Cotton South, if applied to Tennessee, would greatly exaggerate the incidence of sharecropping among the state's freedmen. It would also, as will be seen, significantly understate the extent to which Tennessee blacks acquired and lost land during the Reconstruction era. The second distortion is as critical as the first, particularly given that the consequences of the failure to effect land reform after emancipation have become central to Reconstruction historiography.

Historians investigating the record of black land accumulation after emancipation have typically adopted one of two strategies. Several have attempted to determine the proportion of freedmen owning land at various points in time,[49] whereas others have sought to chart changes over time in

49. See, for example, Joel Williamson, *After Slavery: The Negro in South Carolina during Reconstruction, 1861–1877* (Chapel Hill: University of North Carolina Press, 1965), pp. 144–55; Joe M. Richardson, *The Negro in the Reconstruction of Florida, 1865–1877* (Tallahassee: Florida State University Press, 1965), pp. 73–82; Edward H. Bonekemper III, "Negro Ownership of Real Property in Hampton and Elizabeth City County, Virginia, 1860–1870," *Journal of Negro History* 55 (1970):165–81; Kolchin, *First Freedom*, pp. 134–9, 148; Frank Jackson Huffman Jr., "Old South, New South: Continuity and Change in a Georgia County, 1850–1880" (Ph.D. diss., Yale University, 1974), pp. 127–30; Shifflett, *Patronage and Poverty in the Tobacco South*, pp. 18–20, 52–3; Orville Vernon Burton, *In My Father's House Are Many Mansions: Family and Community in Edgefield, South Carolina* (Chapel Hill: University of North Carolina Press, 1985), p. 260; Barbara Jeanne Fields, *Slavery and Freedom on the Middle Ground: Maryland During the Nineteenth Century* (New Haven: Yale University Press, 1985), pp. 175–9. These efforts have been significantly augmented by the recent impressive contribution of Loren Schweninger, *Black Property Owners in the South, 1790–1915* (Urbana: University of Illinois Press, 1990), especially Chap. 5. Schweninger has examined manuscript census data from every southern state to estimate the proportion of freedmen owning land in 1870. Unfortunately, for post-1870 analysis Schweninger relies on published census figures on the proportion of owners among black farm operators – a

the proportion of total wealth that the freedmen commanded.[50] Both approaches afford important insights into the freedmen's struggle for independent landownership, yet both may obscure as well as reveal. Specifically, both strategies invariably underestimate the extent to which individual freedmen acquired or lost farms of their own over time and exaggerate the degree to which stasis characterized the Reconstruction era. A longitudinal analysis, on the other hand, that traces individual farmers over a period of years affords more precise estimates of individual economic mobility and results in a more balanced view of postbellum social and economic reorganization.

A careful examination of land accumulation among Tennessee freedmen demonstrates this point effectively. Census and tax records show that in the sample counties as a whole 4 percent of black household heads owned their own farms by 1870. During the next ten years, this proportion approximately doubled, rising to 7.9 percent by 1880. The proportion of freedmen owning land increased from 8.7 to 18.1 percent in East Tennessee, from 9.5 to 14.3 percent in Middle Tennessee, and from 1.4 to 4.4 percent in West Tennessee. As indications of overall structural change within the black population, these figures are valuable and precise. As measurements of the extent of individual mobility, however, such statistical "snapshots" leave much to be desired and are most accurate for nonexistent closed communities of freedmen in which no one ever died or moved away and no landowners ever lost their land.

In every section of the state the percentage of persisting freedmen who acquired land during the 1870s far exceeded the percentage-point increase in overall black landownership during the decade (see Figure 4.2). Consid-

category that excluded thousands of blacks who worked the soil as wage laborers. The 1870 estimates, then, are not comparable with those for subsequent years, which exaggerate the proportion of farmowners among black farm households.

50. The three most thorough econometric studies of the postbellum South all base their analyses of black landownership on the same source, Georgia tax data, which were first mined almost ninety years ago by W. E. B. DuBois. See W. E. Burghardt DuBois, "The Negro Landholder of Georgia," *United States Department of Labor Bulletin* No. 35 (Washington, DC: Government Printing Office, 1901), pp. 647–777; also Ransom and Sutch, *One Kind of Freedom*, pp. 83–5; Robert Higgs, *Competition and Coercion: Blacks in the American Economy, 1865–1914* (Chicago: University of Chicago Press, 1977), pp. 69–70; and Wright, *Old South, New South*, p. 106. See also Robert Higgs, "Accumulation of Property by Southern Blacks Before World War I," *American Economic Review* 72 (1982):725–37; and Robert A. Margo, "Accumulation of Property by Southern Blacks Before World War I: Comment and Further Evidence," *American Economic Review* 74 (1984):768–76.

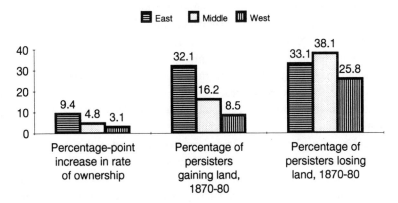

Figure 4.2   Freedmen and landownership, sample Tennessee counties, 1870–80

ering the sample counties as a whole, more than 12 percent of the 1870 landless freedmen who remained in the same county owned land of their own by 1880, even though the overall proportion of freedmen who owned land increased by not quite 4 percentage points. On the other hand, a huge percentage of those rare freedmen who had acquired farms by 1870 were landless by the end of the next decade, a sizable downward trend that the decennial ownership rates fail entirely to reflect.

The latter discovery is particularly significant. Considering the great importance attached to the land question, it is surprising how little scholars have tried to determine whether blacks who successfully acquired farms were also able to hold on to them. The question is a crucial one; the implicit assumption underlying most scholarly criticism of Republican Reconstruction policies has been that land redistribution would have made a lasting difference in the economic (and political) position of the freedmen within southern society. As Figure 4.2 shows, however, in Tennessee between approximately one-fourth and two-fifths of freedmen who owned land in 1870 lost title to their farms within the next ten years; in comparison to white landowners, black owners who persisted through 1880 were 2.5 to 4 times more likely to lose their lands (refer to Table 3.5).

The experience of black landowners in Tennessee during the 1870s clearly casts doubt on the long-term efficacy of a Congressional land grant to the freedmen and lends support to those scholars who have been uncertain whether the freedmen possessed the managerial experience and the capital resources necessary to withstand the economically depressed decades

of the late nineteenth century.[51] Undoubtedly a number of other factors –
including sheer racial prejudice – contributed to the poor showing among
black landowners. Nevertheless, given the high rate of failure among Tennessee blacks during the 1870s, one must question just how many former
slaves would have been able to keep their farms throughout the recurring
depressions of the post–Reconstruction era.[52]

Although tracing individual Tennessee freedmen across time reveals substantially greater mobility than the standard scenario typically allows, it is
important to note the regional variations that obtained within the larger
statewide pattern (see Figure 4.2). As elsewhere in the South, black landownership in Tennessee was considerably more common in nonplantation
than in plantation areas.[53] In particular, persistent freedmen in East Tennessee gained their own farms proportionally twice as often as did blacks
in Middle Tennessee and four times more frequently than did the freedmen
of West Tennessee. In comparison with the percentages of landless freedmen obtaining land, there was far less regional variation in the likelihood
of dispossession; East Tennessee freedmen, for example, were about as
likely to lose their farms as those from other regions.

That black progress should have been markedly greater in East Tennessee is not especially surprising; by 1910, across the South as a whole, blacks
living in nonplantation areas were three times more likely to own land than
were blacks in plantation areas.[54] Why this was so is unclear, however. One
possible explanation can be readily dismissed: It is highly improbable that
freedmen in East Tennessee faced less racist opposition to their advance-

51. Herman Belz, "The New Orthodoxy in Reconstruction Historiography," *Reviews
   in American History* 1 (1973):111–12; Wharton, *The Negro in Mississippi*, p. 60;
   David M. Potter, *Division and the Stresses of Reunion, 1845–1876* (Glenview, IL:
   Scott, Foresman and Company, 1973), pp. 186–7; William Gillette, *Retreat from
   Reconstruction, 1869–1879* (Baton Rouge: Louisiana State University Press, 1979),
   p. xiv; Carl N. Degler, "Rethinking Post–Civil War History," *Virginia Quarterly
   Review* 57 (1981):255.

52. Joseph P. Reidy argues that the freedmen, had they received a Congressional land
   grant, would have forgone staple cultivation for subsistence farming, thus sheltering
   themselves from the vagaries of the postbellum economy. Although a feasible scenario, it is instructive to note that black landowners in East and Middle Tennessee,
   who focused exclusively on the production of foodstuffs, were just as likely to lose
   their land as black farmowners in the cotton counties of West Tennessee. See
   Reidy, "Slavery, Emancipation, and the Capitalist Transformation of Southern
   Agriculture," in Ferleger, ed., *Agriculture and National Development*, p. 236.

53. For evidence of similar patterns in other parts of the South, see Mandle, *The Roots
   of Black Poverty*, pp. 42–3; Schweninger, *Black Property Owners in the South*, esp.
   Chaps. 5 and 6.

54. Mandle, *The Roots of Black Poverty*, pp. 42–3.

ment. State Freedmen's Bureau chief Clinton B. Fisk made this point explicitly in testimony before Congress in 1865, noting the "melancholy fact that among the bitterest opponents of the negro in Tennessee are the intensely radical loyalists of the mountain districts." The same year journalist John Trowbridge relayed an identical finding to northern readers, albeit more succinctly. "East Tennesseans," he reported, "though opposed to slavery and secession, do not like niggers." What Fisk and Trowbridge had come to recognize was that underlying the fervent unionism of most white East Tennesseans was the conviction, as expressed by a Knoxville newspaper, that the prosecession "aristocracy" that had "cursed" the region "was an aristocracy founded alone upon the nigger." The northern journalist Whitelaw Reid also recognized these twin antipathies during his travels in the eastern part of the state. As he observed in retrospect,

> ... it was manifest that East Tennessee radicalism, however earnest on the question of punishing Rebels, did not go to the extent of defending the Negroes. . . . The prejudices against them were, with the most, intense; and if any way of driving them out of the country can be found, it will be very apt to be put into force.[55]

It is likely that the greater success of East Tennessee blacks in acquiring farms was a product of peculiar economic, demographic, and political circumstances that made land more affordable and lessened the hostility of whites to black landownership. On the one hand, because of less fertile soil and greater distance to markets, farm land was typically cheaper there. During the 1870s, the average assessed value of land per acre was $4.83 in East Tennessee, compared to $7.75 in Middle Tennessee and $6.87 in West Tennessee.[56] On the other hand, because slaves had always constituted a tiny proportion of the area's antebellum labor force, the possibility of their transformation into a free black yeomanry, though distasteful to mountain whites, did not threaten to disrupt the local economy.[57] In addition, blacks in heavily Republican East Tennessee – in contrast to freedmen elsewhere

55. Testimony of Clinton B. Fisk in *Report of the Joint Committee on Reconstruction*, pt. I, p. 122; John Trowbridge, *The South* (Hartford, CT: L. Stebbins, 1865), p. 239; *Brownlow's Knoxville Whig and Rebel Ventilator*, 20 April 1864; Whitelaw Reid, *After the War*, p. 352.

56. *Appendix to House and Senate Journals, Tennessee, 1881* (Nashville: Tavel & Howell, 1881), pp. 31–3.

57. Considering farm household heads in the eight sample counties only, landless blacks in 1880 constituted approximately 12 percent of all landless farmers and farm laborers in East Tennessee, 39 percent in Middle Tennessee, and 85 percent in West Tennessee.

in the state – did not constitute a challenge to the political hegemony of local whites. Indeed, mountain Republicans relied heavily on black votes to insure their control of local and Congressional races. As a consequence, they may have been willing to tolerate a degree of black progress that white Democrats in other sections of the state deemed unacceptable. A *New York Times* correspondent suggested as much in 1867, observing that "there was more and bitterer prejudice against the blacks in East Tennessee than elsewhere," but that "when their interest demands it they [mountain whites] forget all these prejudices."[58]

Whether the overall rate of land accumulation among Tennessee blacks was "agonizingly slow" or impressively rapid is primarily a matter of perspective.[59] Given the hopes they had entertained in 1865, many freedmen must have viewed the extent of their progress as bitterly disappointing, and with good reason. During the 1870s they had been far less successful than landless whites in acquiring land. Additionally, with the exception of land-owning freedmen in West Tennessee, the relatively few blacks who successfully purchased farms consistently possessed much smaller and much less valuable plots than those owned and operated by whites. (In West Tennessee, in contrast, although blacks were much less likely to acquire farms, those who did on average owned plots roughly comparable in size to typical white-owned farms.) Considering the sample counties as a whole, black owners farmed units less than half as large and two-thirds less valuable (see Table 4.3). As a consequence, although freedmen comprised one-quarter of the total farm population in 1880, they still owned less than 2 percent of total farm acreage.[60]

Advocates of the standard scenario often present statistics such as the last one to imply that land acquisition by freedmen was so minimal that it

58. *New York Times*, 15 February 1867. See also Gordon B. McKinney, *Southern Mountain Republicans, 1865–1900: Politics and the Appalachian Community* (Chapel Hill: University of North Carolina Press, 1978), especially pp. 132–41.

59. Ransom and Sutch maintain that black progress was "agonizingly slow"; Robert Higgs, paraphrasing Frederick Douglass, observes that "one might well marvel that the freedmen did so well." See Ransom and Sutch, *One Kind of Freedom*, p. 8; Higgs, *Competition and Coercion*, p. 61.

60. Black landowners may have also been less likely than were white owners to own their farms "free and clear" (i.e., their farms may have been more frequently and more heavily mortgaged than were white-owned farms). Unfortunately, it has been impossible to test this hypothesis effectively for the sample Tennessee counties. Property mortgages (most typically in the form of deeds of trust) appear to have been incompletely and erratically recorded by local officials throughout the Reconstruction years.

Table 4.3 Median size and value of owner-operated farms by race
of operator, sample Tennessee counties, 1880

	East	Middle	West
Total acres			
Whites	100	104	128
Blacks	20	34	96
Farm value ($)			
Whites	800	1176	1000
Blacks	200	350	900

Source: Eight-county sample, see text.

may be disregarded when analyzing the reorganization of southern agriculture. This conclusion is unjustified and unfortunate. Of the various barometers of black land accumulation, none is more penetrating and yet potentially more deceptive than the proportion of farm acreage blacks controlled. Unquestionably, the striking disparity between the numerical importance of Tennessee freedmen and their minimal command of wealth is powerful testimony of their persistent overall poverty relative to whites. It would be wrong, however, to infer from such figures that their limited wealth accumulation was inconsequential. To do so would be to misinterpret the significance of those gains, both to the freedmen as well as to their former masters.

To begin with, once the ex-slaves had resigned themselves to a struggle for land without benefit of federal intervention, the undeniable success of a minority during the 1870s may have proven a source of encouragement to the remainder. Even though they owned only 2 percent of the state's farm land, one out of eight persisting freedmen had acquired farms of their own during the decade, proving that landownership was a realistic, albeit formidable, goal. This optimistic influence, for example, may help to explain the significant opposition among Tennessee freedmen to black emigration schemes. Lincoln County freedman R. F. Hurly, for instance, spoke against the efforts of the 1875 Colored Emigration Convention by arguing that "many of our people are doing well in this State."[61]

Furthermore, to former masters desiring an abundant, stable labor force, the accomplishments of black landowners, however rare, constituted an

61. *Nashville Union and American*, 21 May 1875.

ominous threat that far transcended the economic significance of their land-holdings. If historians have interpreted correctly the fears of southern whites, then what former masters feared most, at least after the threat of wholesale land redistribution had passed, was not that blacks would amass agricultural wealth in quantities proportional to their numbers. On the contrary, they were concerned that blacks would acquire tiny plots sufficiently large for subsistence farming, thus removing themselves from the pool of laborers available for service to white employers.[62] As a correspondent to the *Southern Cultivator* complained, blacks who owned merely a "patch of ground" would "almost starve and go naked before they [would] work for a white man."[63] By focusing on the small proportion of wealth owned by freedmen – rather than the frequency with which they acquired farms, however small – scholars inadvertently minimize the significant challenge that even limited accumulation posed to the social and economic order.

Conclusion

The results of this investigation raise important questions for historians of the Reconstruction era. Although it is possible that Tennessee's experience was unique among southern states, or characteristic of the Upper South only, there are no *a priori* reasons to assume this was so. Students of the Cotton South generally have not undertaken the painstaking analyses necessary to establish the validity of the standard scenario. Indeed, two of the most pressing needs in the field of postbellum southern history are for broad-based, systematic analyses that determine the continued extent and function of black farm laborers (not operators) and longitudinal analyses that follow across time the efforts of individual freedmen to acquire and retain land.[64]

62. See, for example, Eric Foner, *Nothing But Freedom: Emancipation and Its Legacy* (Baton Rouge: Louisiana State University Press, 1983), Chap. I; and Jaynes, *Branches Without Roots*, Chap. IV.
63. *Southern Cultivator* 25 (1867):67.
64. Significantly, those few scholars who have tried to assess the prevalence of farm laborers during the postbellum era have consistently concluded that among blacks laborers outnumbered sharecroppers and tenants combined. See Hahn, *The Roots of Southern Populism*, p. 157; Charles L. Flynn Jr., *White Land, Black Labor: Caste and Class in Late-Nineteenth Century Georgia* (Baton Rouge: Louisiana State University Press, 1983), pp. 66–8; David F. Weiman, "Petty Commodity Production in the Cotton South: Upcountry Farmers in the Georgia Cotton Economy, 1840–1880" (Ph.D. diss., Stanford University, 1983), p. 417; Davis, *Good and Faithful*

Exploration of this sort has proven that with regard to Tennessee the standard scenario regarding the freedmen and the post-emancipation transformation of southern agriculture is factually incorrect in two fundamental respects. First, the institutional reorganization of agriculture in the state was neither swift nor thorough, nor did it result in the immediate predominance of sharecropping among the former slaves. Between 1860 and 1880 the number and average size of farm units across the state underwent major changes, but to a surprising degree these reflected a remarkable increase in the number of *white operators*. Although sharecropping and tenancy did grow in importance, as late as 1880 the typical freedman was more likely to have been a wage laborer than a cropper or tenant.

Second, despite the continued concentration of blacks at the very lowest rung of the agricultural ladder, in Tennessee there was considerable fluidity between the landholding and landless ranks. Throughout the 1870s a small but significant proportion of former slaves purchased farms of their own; at the same time, however, a substantial fraction of those who began the decade as owners eventually lost title to their farms, testimony to the fragility of black landownership during the Reconstruction era.

A primary strength of the standard scenario of postbellum reorganization has been its emphasis on the chaotic disorder of the immediate postbellum years; perhaps its greatest flaw has been the tendency to overlook continued experimentation, flux, and uncertainty thereafter. Certainly the white Tennesseans who complained to their commissioner of agriculture in the 1880s regarding the reliability of black labor would not have understood modern scholarly descriptions of their world that stress stability and order. Only fifteen years removed from slavery, their former slaves continued to move across the countryside with a frequency irritatingly similar to that of landless whites. Under the labor arrangements that were emerging from such apparent chaos, one-half of their former slaves worked as wage laborers. The other half consisted of an amalgam of sharecroppers, tenants, and, from their point of view, a disturbingly large fraction of potentially independent owner-operators. If a new system had indeed crystalized in Tennessee agriculture by 1880, it bore little resemblance to the antebellum order.

Labor, pp. 164–5; Irwin, "Farmers and Laborers," pp. 53–60; Kinzer, *Kinship and Neighborhood in a Southern Community*, p. 112.

5. Agricultural Change to 1880

An arresting juxtaposition of continuity and change characterized the lives of Tennessee's white farm population during the Civil War era. Minor variations aside, antebellum patterns of wealth distribution and economic mobility survived the war to an impressive degree, strikingly resistant to the staggering economic losses that the war inflicted. Socioeconomic stability among whites coincided with revolutionary change between the races, however. Although the war had resulted neither in the downfall of the prewar elite nor in the proletarianization of the agricultural middle class, it had undeniably effected the "sudden and violent overthrow" of the southern labor system. With varying degrees of optimism and dismay, Tennesseans all across the state assumed that emancipation would lead to radical alterations in their agricultural economy.[1] Keeping in mind that post-emancipation reorganization was an ongoing process, this chapter analyzes and compares changes in farm operations that had occurred by 1880. To highlight differences from antebellum patterns, it focuses on several of the same factors stressed in Chapter 1: the scale of farm operations, the structure of the farm community, the degree of self-sufficiency among individual farmers, and the level of income that farming afforded. The analysis should help not only to assess the extent of discontinuity during the Civil War era but also to evaluate the degree to which agricultural diversity divided the state during the early postbellum years.

1. See, for example, *Southern Cultivator* 27 (Feb. 1869):54; Hermann Bokum, *The Tennessee Handbook and Immigrant's Guide* (Philadelphia: J. B. Lippincott & Co., 1868), pp. 118–19; Joseph Buckner Killebrew, *Introduction to the Resources of Tennessee* (Nashville: Tavel, Eastman, & Howell, 1874; reprint ed., Spartanburg, SC: The Reprint Company, 1974), p. 350.

Scale of Operations

"The New South," Sidney Lanier proclaimed in an essay at the end of the nineteenth century, "means small farming." One of the most conspicuous developments in southern agriculture after emancipation was the rapid disappearance of the plantation system of production (though not of large landholdings) and a drastic, widespread decline in the scale of agricultural operations. Among the former states of the Confederacy, the number of farm units, including plots worked by tenants or sharecroppers, increased by nearly 140 percent between 1860 and 1880, while mean farm size dropped by nearly three-fifths, falling from 365 to 157 acres. Every ex-Confederate state was affected profoundly; the proportional decline in average farm size was just slightly less precipitous in the Upper South (52 percent) than in the Lower South (62 percent). Tennessee was actually one of the least affected; as noted in the preceding chapter, the number of farm units across the state roughly doubled while the average size of farms fell almost precisely by half. Although the process took time to unfold, the trend itself was almost immediately apparent. "One of the points I have seen," a large Middle Tennessee landholder reported to an absent relative less than two years after Appomattox, "is the tendency of farming to be done on a small scale. . . . I think therefore the amount we will cultivate (not graze) will tend to grow smaller."[2]

As one would expect, in Tennessee the decline in farm size was greatest in the southwestern corner of the state, the area where large plantations had been most prevalent during the antebellum period (see Table 5.1 and Figures 5.1a–c). In Fayette and Haywood counties a nearly fourfold increase in farm units between 1860 and 1880 caused median farm size to plummet by two-thirds. The proportion of farms with two hundred acres or more fell from over one in three on the eve of the Civil War to only one in twenty by 1880. Conversely, the proportion of very small farms (fewer than fifty improved acres) tripled during these years, so that by the

2. Sidney Lanier, "The New South," in *Retrospects and Prospects* (New York: C. Scribner's Sons, 1899), p. 104; R. S. Ewell to Major Campbell Brown, 10 March 1867, box #1, file #13, Brown–Ewell Papers, Tennessee State Library and Archives. Estimates of farm size are derived from United States Census Office, Tenth Census [1880], *Report on the Production of Agriculture in the United States at the Tenth Census* (Washington, DC: Government Printing Office, 1883), p. 25. Following convention, I define the Upper South to include Arkansas, North Carolina, Tennessee, and Virginia, but I also include West Virginia because of the difficulty of disaggregating that state from the rest of Virginia in calculating antebellum figures.

Table 5.1 Farm size profile, sample Tennessee counties, 1860-1880

	East	Middle	West
Mean farm size in improved acres			
1860	79.8	108.2	206.1
1880	55.0	63.0	56.9
Median farm size in improved acres			
1860	70.0	75.0	100.0
1880	38.0	41.0	34.0
Percentage of farms in _1880_ with:			
1-49 improved acres	59.4	54.8	66.0
50-199 improved acres	37.6	40.4	29.1
200+ improved acres	3.1	4.8	5.0

Source: Eight-county sample, see text.

end of the 1870s fully two-thirds of all farm units fell into this category. By 1880, then, if not earlier, the plantation system of production had collapsed and small farms dominated the landscape of the West Tennessee Black Belt.

To a surprising extent, the rest of the state also shared in this tendency toward smaller farms, albeit less dramatically. In the mixed-farming regions of East and Middle Tennessee, median farm size dropped by nearly one-half during the twenty years after 1860. Although large farms or plantations had been comparatively rare in both sections, the size distribution of farms underwent a thoroughgoing transformation in each. In East Tennessee, for example, where slaves accounted for less than one-tenth of the prewar population and fewer than one in 500 antebellum farmers qualified as a planter, the proportion of farms with fewer than fifty improved acres soared from approximately 30 to 59 percent, a trend roughly duplicated in the central part of the state. Truly large farms, in contrast, had very nearly disappeared from both sections.

The combined effect of these changes was to eliminate entirely the disparity in farm size that had distinguished the state's major regions before emancipation. Slavery had provided an elastic supply of labor to individual farmers and allowed an indefinite expansion of farm size. Consequently, as

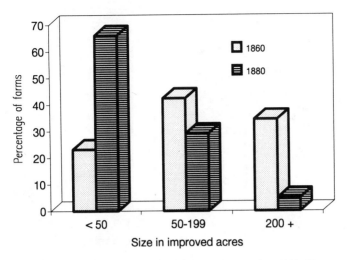

Figure 5.1a Farm size profile, sample West Tennessee counties, 1860–80

one moved westward across the state in 1860 the size distribution of farms had been increasingly skewed in favor of large farms. This was reflected in the *mean* size of farms, which had been 2 to 2.5 times higher in the major slaveholding counties of West Tennessee than elsewhere in the state. Emancipation undermined the labor advantage of the former slaveholders, however, so that by 1880 mean farm size varied insignificantly from one end of the state to the other.

The typical scale of farm operations, of course, had never differed as greatly among the regions as mean figures had seemed to suggest. *Median* farm size in West Tennessee in 1860 had been only 43 percent greater than in East Tennessee and 33 percent greater than in Middle Tennessee. These smaller differences also disappeared during the aftershocks of the Civil War. Within fifteen years of the war's end, the major divisions of the state varied negligibly. Notably, by 1880 median farm size was actually smallest in West Tennessee. Overall, more than one-half of the farms across the eight sample counties consisted of fewer than forty improved acres. In an unrelenting trend from the Appalachians to the Mississippi, small farms had become the hallmark of Tennessee agriculture.

The Changing Structure of the Farm Population

By integrating the previous two chapters' separate analyses of the black and white farm populations, it is possible to recreate with some confidence a

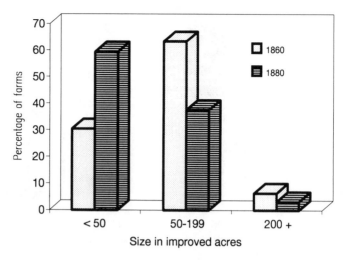

Figure 5.1b   Farm size profile, sample East Tennessee counties, 1860–80

post-emancipation scenario that explains the extraordinary alterations in scale that have been outlined. To begin with a factor seldom considered, population pressures in all three regions of the state placed considerable strain on local supplies of arable land. The *white* farm population in Tennessee as a whole increased by two-fifths between 1860 and 1880, during which time total farm acreage in the state was almost literally unchanged.[3] In such circumstances some decline in farm size would appear to have been inevitable. Indeed, in both East and Middle Tennessee the pressure of population growth on finite resources of land likely outweighed emancipation as a factor promoting the declining scale of farm operations. This is particularly surprising with regard to Middle Tennessee, where nearly one-third of the antebellum population had been enslaved.

Considering the common focus among scholars on the growth of post-bellum tenancy, it is important to note that the impressive overall growth of the white agricultural population after 1860 rested to a significant degree on expansion in the ranks of white farm*owners*. Although the number of white-owned farms increased more slowly than the number of white-operated farms, the increase was impressive nevertheless. In the eight sample counties, for instance, the number of white-owned farms increased by 46 percent between 1860 and 1880. Because improved acreage

3. Across the state as whole, total farm acreage decreased between 1860 and 1880 by fewer than 3,000 acres from an original total of nearly 21 million. In the eight sample counties total acreage declined by 0.02 percent.

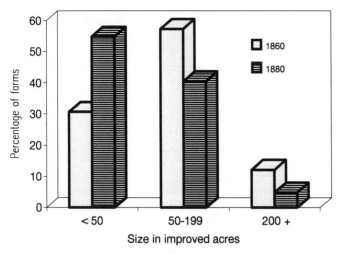

Figure 5.1c Farm size profile, sample Middle Tennessee counties, 1860–80

in the sample counties increased by less than 8 percent, the growth of white farmownership clearly influenced the declining scale of agricultural operations.

In every part of the state, of course, the demise of slavery also contributed substantially to falling farm size. In Tennessee, as elsewhere across the South, emancipation sent shock waves throughout the countryside by destroying the labor system on which much of agricultural production had been founded. The result, in the short run anyway, was a widespread shortage of labor that varied across the state in relation to the prewar importance of slavery to the local economy. From the outset, complaints among white landowners concerning a deficiency of workers were probably heightened by an aversion to contracting with their former slaves and by a conviction that free black labor was inherently unreliable. Even so, there was more to this shortage than simply the distorted perceptions of racist ex-masters. The well-documented restlessness of the freedmen, particularly their tendency to move from the farm to the city, as well as the loss during the war of perhaps one-fifth of adult white males combined to create a scarcity of labor that was a real and severe phenomenon. Clinton B. Fisk, chief officer of the Freedmen's Bureau in Tennessee and Kentucky, noted as much in testimony before the Joint Committee on Reconstruction in 1866. In response to a query regarding the proportion of freedmen who were able to find jobs, Fisk replied, "All of them in Tennessee who can do any work.

Table 5.2 Estimated tenancy rates among Tennessee farm operators, 1860-1880 (percentages)

	East	Middle	West
All operators, 1860	16.0	21.4	13.2
<u>White</u> operators only, 1880	22.5	24.5	32.6
All operators, 1880	23.6	29.2	63.5

Source: Eight-county sample, see text.
Note: Tenants in 1860 included all farm operators who appeared on the agricultural census schedule but reported no real wealth on the population schedule. Tenants in 1880 consisted of all individuals listed on the agricultural schedule as farming for either a fixed rent or a share of the crop.

I could furnish employers for 25,000 more laborers than I have, such is the demand for labor in the valley of the Mississippi."[4]

In differing degrees of intensity across the state, this shortage of workers prompted a period of experimentation among landowners, who tested a variety of strategies aimed at attracting and maintaining a sufficient supply of agricultural labor. Seeking to preserve as much continuity as possible with the *status quo ante bellum*, most landowners in need of labor first sought to hire white or black workers as wage employees, paying either a set amount of cash or a stipulated fraction of the farm's output. In so doing, they turned to a relatively familiar prewar arrangement, one that had encompassed a far larger proportion of antebellum whites than historians have heretofore recognized. Increasingly during this period of testing, landowners also turned to some form of agricultural tenancy or sharecropping. For reasons that are examined later, by 1880 large landowners all across the state commonly subdivided their farms into tenant plots farmed by freedmen or landless whites. The result was an increase in the tenancy rate among farm operators that was substantial in the eastern and central sample counties and little short of spectacular in the westernmost section of the state (see Table 5.2).

Although opportunities for freedmen to rent or sharecrop land became increasingly common during the Reconstruction era, most former slaves apparently remained at the very bottom of the agricultural ladder, working

4. United States Congress, *Report of the Joint Committee on Reconstruction*, 39th Cong., 1st sess. (Washington, DC: Government Printing Office, 1866), pt. I, p. 112. For additional evidence of a significant labor shortage see Freedmen's Bureau reports cited in Chapter 4, fn. 32.

Table 5.3a Cross-tabulation of farm households in **East** Tennessee by race and tenure of head of household, 1880

	White	Black	Total
Owner-operator	4,037 (61.2%)	79 (18.1%)	4,116 (58.5%)
Rents for fixed Money Rental	111 (1.7%)	3 (0.7%)	114 (1.6%)
Rents for shares of products	1,056 (16.0%)	105 (24.0%)	1,161 (16.5%)
Not on agricultural schedule	1,396 (21.2%)	250 (57.2%)	1,646 (23.4%)

Source: Eight-county sample, see text.
Note: Column percentages are in parentheses.

by the day, month, or year as wage laborers on farms operated by others (see Tables 5.3a–c). Opportunities for landless whites to rent land, on the other hand, seem to have opened up after emancipation, at least in East and West Tennessee. In both of those sections significant increases in tenancy among white operators after 1860 signified less the downward mobility of smallholders than the shift among landless whites from wage labor to tenancy.

These findings are fundamentally inconsistent with most scholarly assessments of the racial balance of power during the early postbellum period. Most accounts of the transformation of the southern labor system after emancipation posit a confrontation between white landowners determined to implement a gang-labor wage system and freedmen resolved to resist any arrangement reminiscent of slavery. Because of the extreme shortage of labor, it is widely agreed, the uncompromising opposition of the freedmen forced reluctant whites to capitulate and subdivide their farms and plantations among black tenants and sharecroppers. Given this line of argument, the presumed preeminence of sharecropping among the freedmen becomes testimony to the market power of the former slaves and – however dubious in retrospect – represents their one undeniable economic victory in the postwar world.[5]

5. For examples, see Roger Ransom and Richard Sutch, *One Kind of Freedom: The Economic Consequences of Emancipation* (Cambridge: Cambridge University Press, 1977), pp. 56–70; Forrest McDonald and Grady McWhiney, "The South from

Table 5.3b Cross-tabulation of farm households in **Middle** Tennessee by race and tenure of head of household, 1880

	White	Black	Total
Owner-operator	5,597 (61.6%)	379 (14.3%)	5,976 (50.9%)
Rents for fixed money rental	683 (7.5%)	168 (6.3%)	851 (7.2%)
Rents for shares of products	1,133 (12.5%)	486 (18.3%)	1,619 (13.8%)
Not on agricultural schedule	1,677 (18.4%)	1,617 (61.0%)	3,294 (28.1%)

Source: Eight-county sample, see text.
Note: Column percentages are in parentheses.

The discovery that as late as 1880 a majority of freedmen had not yet attained the "elevated" rank of sharecropper or tenant raises serious doubts about the market power of the ex-slaves after the first few years of freedom. In Tennessee, it appears that the agricultural labor shortage caused by war and emancipation had virtually disappeared within fifteen years, if not much sooner, primarily due to the rapid growth of the local farm populations, both white and black. Whereas improved acreage increased by only 6 to 8 percent in the sample counties between 1860 and 1880, the total farm population grew by 43 percent in the eastern counties, 38 percent in the central counties, and 56 percent in the southwestern cotton counties. To the extent that freedmen ceased to work like slaves, the latter figures must be revised downward to reflect the true change in the growth of agricultural *labor* (as opposed to the agricultural population). There is abundant evidence that the output of black labor per capita declined after emancipation as the work day shortened and the rate of participation in the labor force declined noticeably, especially among women and children. Roger Ransom and Richard Sutch have estimated the decline among blacks

Self-Sufficiency to Peonage: An Interpretation," *American Historical Review* 85 (1980):1115; Ronald L. F. Davis, *Good and Faithful Labor: From Slavery to Share-cropping in the Natchez District, 1860–1890* (Westport, CT: Greenwood Press, 1982), p. 190; Michael Wayne, *The Reshaping of Plantation Society: The Natchez District, 1860–1880* (Baton Rouge: Louisiana State University Press, 1983), pp. 132–40.

Table 5.3c Cross-tabulation of farm households in **West** Tennessee by
race and tenure of head of household, 1880

	White	Black	Total
Owner-operators	1,921 (65.6%)	265 (4.4%)	2,186 (24.5%)
Rents for fixed money rental	647 (22.1%)	1,443 (24.1%)	2,090 (23.4%)
Rents for shares of products	281 (9.6%)	1,428 (23.8%)	1,709 (19.2%)
Not on agricultural schedule	79 (2.7%)	2,860 (47.7%)	2,939 (32.9%)

Source: Eight-county sample, see text.
Note: Column percentages are in parentheses.

in the Cotton South at between 28 and 37 percent per capita.[6] Even adopting their upper bound (and assuming constant man-hours per capita among
whites), it can be shown that the total supply of agricultural labor in man-
hours increased by 39.9 percent in the sample East Tennessee counties,
26.5 percent in the Middle Tennessee counties, and 17.5 percent in the
West Tennessee counties. In every region of the state, in other words, labor
was relatively more plentiful in 1880 than it had been at the end of the
antebellum era. This does not mean that competition for workers had
ceased to be an important factor in the rural labor market but simply that
white landowners were clearly not at the mercy of their former slaves to
the extent often maintained.[7]

With this in mind, the complaints among white Tennesseans regarding a
shortage of labor well into the 1870s are difficult to take literally. It is not the
implied distress that was artificial but its imputed cause.[8] White landowners
suffered less from a shortage of labor *per se* – that is, a deficiency of willing,

6. Ransom and Sutch, *One Kind of Freedom*, pp. 44–7.
7. For example, in explaining the rise of sharecropping in the Natchez District of
   Louisiana and Mississippi, Ronald Davis argues that white landowners "had no
   choice in the matter," and "literally were dragged kicking and screaming into the
   system." See *Good and Faithful Labor*, p. 190.
8. Strictly speaking, a true shortage of labor may be said to exist whenever the average
   wage offered by employers falls below a competitive equilibrium level. It goes
   without saying that this was not what white landowners had in mind when voicing
   their complaints.

able-bodied workers – than from a shortage of reliable labor – that is, workers who resembled slaves in their perceived loyalty and dependability. As a correspondent to the *Southern Cultivator* explained as early as the summer of 1870, it was not so much that "muscle" was scarce, but that "the power to control it [was] lacking." A county-level survey of labor conditions taken in the mid-1870s by the state Bureau of Agriculture repeatedly made the same distinction, albeit less explicitly. In Benton County, for example, the survey discovered "a fair demand for *good* hands, which, at present, are very scarce." The reason for the "scarcity" was clear:

> The principal available labor now is negro labor, which is very unreliable; there are, however, some white laborers, and they are generally regarded as being very reliable. The people would be glad to welcome white men who are willing to work for wages, but they insist that they don't want any fresh installments of negroes.[9]

This enduring aversion to free black labor may help to explain the marked structural shift among landless whites in West Tennessee from wage labor to tenancy after emancipation. As late as 1860 nearly three-fifths of landless whites in Fayette and Haywood counties were not attached to any particular farm and probably hired out for wages. By 1880, despite a sizable influx of additional landless whites into the region, more than nine-tenths were tenants or sharecroppers. Actually, the Bureau of Agriculture survey indicates that contemporary impressions of the reliability of white labor were somewhat mixed. Blacks were probably considered more pliable laborers, whites more likely to have the managerial abilities necessary for independent operations. Mixed reviews aside, on the whole it appears that "white men who [were] not afraid to work" were considered eminently preferable to the freedmen.[10] Compared to the ex-slaves, however, landless whites in West Tennessee were indeed scarce; before emancipation adult male slaves had outnumbered white wage laborers by approximately 14:1. Consequently, landowners who sought to attract white labor probably were forced regularly to proffer the inducement of a tenancy arrangement rather than

9. *Southern Cultivator* 28 (1870):198; J. B. Killebrew, *Introduction to the Resources of Tennessee* (Nashville: Tavel, Eastman, and Howell, 1874), p. 1027, emphasis added. See also idem, pp. 1034, 1042, 1051, 1059, 1074, 1083, 1108.
10. Killebrew, *Introduction to the Resources of Tennessee*, p. 1042. For positive views of white labor, see also pp. 1027, 1083, 1108, 1146, 1190; for critical evaluations, see pp. 1034, 1051, 1066, 1100, 1129. For negative assessments of laborers of both races that nevertheless deem whites to be preferable, see pp. 1051, 1074.

a labor contract. If any group had extreme market power in postbellum West Tennessee, it was the landless whites, not the freedmen.

In sum, Tennessee agriculture did undergo a revolution in scale during the Civil War era but for reasons different from those commonly offered to explain similar developments in the Deep South. Without comparable analyses of other southern states, it is impossible to know whether developments in Tennessee were mirrored elsewhere; nevertheless, the findings presented here prompt two observations that historians of southern agriculture and race relations may wish to consider. First, the figures for West Tennessee demonstrate just how deceptive tenancy rates and farm size statistics can be as indicators of the rise of black tenancy and sharecropping. In Fayette and Haywood counties the number of farm units nearly quadrupled between 1860 and 1880, mean farm size fell by three-fourths, the tenancy rate skyrocketed from 13 to 64 percent, and yet one-half of the freedmen in all probability still hired out for wages. The slave-style system of labor had most certainly been obliterated, yet tenancy and wages had emerged as roughly equivalent successors in terms of the numbers of freedmen that they encompassed.

Second, although a more speculative point, it is worth noting that the relatively modest success of Tennessee freedmen in becoming farm operators is far more consistent with other aspects of the black economic experience after the Civil War than scenarios that interpret an assumed preeminence of tenancy or sharecropping among the freedmen as testimony of African-American strength. Most accounts of the fate of the freedmen in postbellum southern agriculture exhibit a troubling dualism, on the one hand describing their striking success in rejecting the imposition of wage labor, on the other hand chronicling their apparent impotence in the face of repression, intimidation, or systematic discrimination with regard to geographic mobility, access to credit, or the acquisition of land. This dualism is patently inappropriate with regard to the freedmen in postbellum Tennessee, who were both less successful at avoiding wage labor and more successful at acquiring land than the standard scenario would allow.

The Growing Importance of Tenancy

The continued importance of wage labor among the freedmen and, to a lesser degree, among landless whites, should not obscure the fact that tenancy and sharecropping became substantially more important to the state's agricultural economy after the Civil War (see Table 5.4). Change was by

Table 5.4 The growing importance of tenancy in Tennessee agriculture, 1860-1880

	East	Middle	West
Percentage of improved acreage on tenant units			
1860	11.7	7.7	5.0
1880	16.1	19.7	41.2
Percentage of farm production produced by tenants[a]			
1860	10.1	9.6	5.7
1880	17.2	22.6	46.3

[a]Based on the estimated value of production in dollars. For method of estimation see Appendix C.
Source: Eight-county sample, see text.
Note: Tenants defined to include all individuals listed in the agricultural schedules as "rent[ing] for fixed money rental" or "rent[ing] for shares of products."

far the greatest in the Black Belt counties of southwestern Tennessee. Before emancipation, tenants and croppers had played a relatively insignificant role in the cotton counties, farming only about 5 percent of the region's improved acreage in 1860 and contributing about the same proportion of agricultural output. By 1880 a revolution had undeniably occurred, as tenants and sharecroppers worked more than two-fifths of all improved acreage and produced nearly one-half of all farm output. Understandably, change was not nearly so pronounced in the mixed-farming sections of eastern and central Tennessee, yet tenancy had increased considerably in importance in those areas as well. The percentage of agricultural output produced on rented units, for example, increased by 70 percent in East Tennessee and by 135 percent in Middle Tennessee.

Because Tennessee landowners who turned to tenancy to meet their labor needs were apparently not forced to do so, their decision to rely on tenant labor is all the more intriguing. Certainly popular opinion did not support their course of action. Contemporaries who commented on the increase in tenancy were almost uniformly critical. A report of the U.S. Department of Agriculture in 1876 characterized the sharecropping plan as "the best possible to destroy fertility and demoralize labor." An agricultural society in South Carolina concurred, maintaining that the system led in-

evitably "to decay and ruin in the farm, and a certain decline in the productive resources of the country."[11]

A recurrent theme among critics concerned the unwillingness of tenants
and croppers to engage in long-term improvements because of their uncertain and tenuous ties to the land. In 1874, for example, Tennessee's
Commissioner of Agriculture raised the oft-repeated complaint that croppers would not work "outside the present crop." On the typical sharecropped farm, as he described it, "fences rotted down, noxious weeds and
shrubs grew without limitation over the farm, and stock-raising became a
thing of the past." Another West Tennessee farmer observed succinctly:
"Whites lose by farming on shares. Negroes are unwilling to manure the
land because of extra work."[12]

The criticism of contemporaries was seconded by the analysis of late-
nineteenth-century scholars. Alfred Marshall, for example, applied the new
concept of marginal analysis to the issue of tenant productivity. His analysis
suggested that, although all tenants would tend to shun long-term improvements, share tenants and sharecroppers would likely stint their input of
daily labor as well. For example, Marshall reasoned that because the sharecropper usually received only one-half of the marginal return to his labor,
he would maximize his income by ceasing additional labor input when the
marginal return was double the marginal cost. The same argument could
be applied to the share tenant.[13]

If the rapid growth of tenancy after emancipation was economically unwise (a point not yet established empirically), and if, furthermore, contemporaries generally believed that this was the case, then landowners must

11. United States Department of Agriculture, *Report of the Commissioner . . . for the
    Year 1876* (Washington, DC: Government Printing Office, 1877), p. 131; *Southern
    Cultivator* 27 (1869):54.
12. Killebrew, *Introduction to the Resources of Tennessee*, p. 351; Eugene W. Hilgard
    [Special Agent], *Report on Cotton Production in the United States* (Washington:
    GPO, 1884), p. 476. See also Robert Somers, *The Southern States Since the War,
    1870–1871* (New York: Macmillan Co., 1871; reprint ed., University: University
    of Alabama Press, 1965), p. 60; Southerner [pseud.], "Agricultural Labor at the
    South," *Galaxy* 12 (1871):330–2; and W. J. Spillman and E. A. Goldenweiser,
    "Farm Tenantry in the United States," in United States Department of Agricul-
    ture, *Agricultural Yearbook, 1916* (Washington, DC: Government Printing Office,
    1917), p. 343.
13. Alfred Marshall, *Principles of Economics*, vol. I, 2d ed. (London: Macmillan & Co.,
    1891), pp. 686–7. See also Rainer Schickele, "Effect of Tenure Systems on Agri-
    cultural Efficiency," *Journal of Farm Economics* 23 (1941):185–207; and Joseph D.
    Reid, "Sharecropping in History and Theory," *Agricultural History* 49 (1975):426–
    40.

have turned to tenant labor for some reason other than economic efficiency. The range of potential motives is extensive. Some may have subdivided their farms or plantations in order to shift part of the burden of risk to tenants or croppers, others perhaps to distance themselves from a close supervision of free blacks that they found distasteful.

Many also probably rented out plots as an extra inducement to attract superior workers. There seems little doubt that landless workers of both races would have preferred any form of tenancy, including sharecropping, to wage labor. Tenant income was undoubtedly higher than the hired hand's, and the high cotton prices of the early postbellum years must have made renting all the more appealing. The tenant was also perceived as more independent than the wage laborer. As a white farmer from Middle Tennessee observed concerning the freedmen, "most negroes are averse to hiring for wages, because of an idea that croppers have greater privileges."[14]

Although the shortage of labor was not sufficient to force landowners to rent to *any* worker who so demanded, the perceived deficiency of reliable labor may well have prompted owners to employ tenant contracts as a means of enticing workers deemed to be especially desirable. Although "desirability" was in the eye of the beholder and no doubt varied from landlord to landlord, it probably included, in addition to an adequate knowledge of farming, character traits such as diligence and deference to authority.[15] Although the prospective white tenant occasionally may have received the benefit of the doubt, the former slave undoubtedly had to earn a reputation for such qualities on an individual basis. This required, above all else, his staying in one place long enough to become well known to local landholders. Considering the significant geographical mobility the freedmen exhibited throughout the 1870s (see Table 5.1), such reputations must have been far from common and, as a result, probably commanded preferential treatment.[16]

In contrast, some historians argue that landlords turned to tenancy not

14. Hilgard, *Report on Cotton Production in the United States*, p. 476; see also Benjamin Hibbard, "Tenancy in the Southern States," *Quarterly Journal of Economics* 27 (1913):485–6.
15. Southerner, "Agricultural Labor at the South," p. 334.
16. On this point see Arthur Raper, *Preface to Peasantry* (Chapel Hill: University of North Carolina Press, 1936), p. 122; Roger Ransom and Richard Sutch, "The Ex-Slave in the Postbellum South: A Study of the Economic Impact of Racism in a Market Environment," *Journal of Economic History* 33 (1973):131–48; Gavin Wright, *Old South, New South: Revolutions in the Southern Economy since the Civil War* (New York: Basic Books, Inc., 1986), pp. 100–1; and Davis, *Good and Faithful Labor*, p. 9.

as a concession to superior workers but as an alternative mechanism by which to perpetuate their control of black labor. Marxist historian Jonathan Wiener has been the most forceful advocate of this point of view. "The origins of sharecropping," Wiener argues, "lie in class conflict. . . . Share-cropping was established as a repressive system of labor allocation and control." Other prominent scholars also emphasize the repressive aspects of the new labor system, arguing that blacks were "confined" within it, "economically enslaved," or "in a position tantamount to debt peonage." In general, those who take this position begin with the presupposition that tenancy was economically inefficient and from this infer that noneconomic motives necessarily shaped the institution's dramatic growth after emancipation.[17]

The recent analyses of cliometricians such as Robert Higgs, Stephen J. Decanio, and Joseph D. Reid directly attack this characterization of tenancy. They argue that the institution was economically efficient and that its rapid growth in the postwar world was the product of economically rational decision making – in Reid's words, "an understandable market response."[18] Among the three, Reid has probably been the most vigorous proponent of the efficiency of tenancy, especially sharecropping.[19] Echoing

17. Jonathan M. Wiener, *Social Origins of the New South: Alabama, 1860–1885* (Baton Rouge: Louisiana State University Press, 1978), p. 69; Jay R. Mandle, *The Roots of Black Poverty: The Southern Plantation Economy After the Civil War* (Durham, NC: Duke University Press, 1978), p. 27; Crandall Shifflett, *Patronage and Poverty in the Tobacco South: Louisa County, Virginia, 1860–1900* (Knoxville: University of Tennessee Press, 1982), p. 25; Steven Hahn, *The Roots of Southern Populism: Yeoman Farmers and the Transformation of the Georgia Upcountry, 1850–1890* (New York: Oxford University Press, 1983), p. 186. See also Gilbert C. Fite, *Cotton Fields No More: Southern Agriculture, 1865–1980* (Lexington: University of Kentucky Press, 1984), p. 5.

18. Joseph D. Reid, "Sharecropping as an Understandable Market Response – The Postbellum South," *Journal of Economic History* 33 (1973):106–30. See also idem, "Sharecropping in History and Theory"; idem, "The Evaluation and Implications of Southern Tenancy," *Agricultural History* 53 (1979):153–69; idem, "White Land, Black Labor, and Agricultural Stagnation: The Causes and Effects of Sharecropping in the Postbellum South," *Explorations in Economic History* 16 (1979):31–55; Robert Higgs, *Competition and Coercion: Blacks in the American Economy, 1865–1914* (Chicago: University of Chicago Press, 1977), pp. 66–8, 138–42; idem, "Race, Tenure, and Resource Allocation in Southern Agriculture, 1910," *Journal of Economic History* 33 (1973):149–69; Lee J. Alston and Robert Higgs, "Contractual Mix in Southern Agriculture since the Civil War: Facts, Hypotheses, and Tests," *Journal of Economic History* 42 (1982):327–53; Stephen J. Decanio, *Agriculture in the Postbellum South: The Economics of Production and Supply* (Cambridge, MA: MIT Press, 1974).

19. Reid generally does not distinguish between share tenants and sharecroppers in his

nineteenth-century critics of the institution, Reid agrees that both fixed-rent tenants and sharecroppers would have shirked long-term improvements and soil conservation practices had landlords allowed them to do so. Most tenant contracts, however, contained provisions outlining specific responsibilities for long-term land improvements, as well as penalties for noncompliance.

Reid also disagrees that the sharecropping system promoted short-run inefficiency. Indeed, the central feature of the system "was the continuing interest of both landlord and tenant in the efficiency of agricultural production." It was especially superior to alternative arrangements in the degree to which it enabled the reduction of "sequential uncertainty," changes that took place over time and allowed time for response. For example, Reid argues that changes in anticipated crop prices or unexpected weather conditions frequently made a renegotiation of crop mix desirable. Because any additional return would be shared (in contrast to fixed-rent tenancy, for instance), both sharecropper and landlord had an incentive to renegotiate. Indeed, Reid goes so far as to maintain that "the rise of tenancy should have increased southern agricultural productivity."[20]

Although his argument is sound theoretically, Reid produces little empirical evidence to support his hypothesis, and the image that it invokes of freedmen and landlords periodically fine-tuning their production strategy in order to take advantage of the latest market reports is a bit difficult to swallow. His contention that the postbellum growth of tenancy should have actually enhanced southern agricultural efficiency is a bold overstatement that, unfortunately, threatens to obscure the more fundamental and more readily defensible argument that he, Higgs, and Decanio are trying to make: The dramatic growth of tenancy after the Civil War is not *a priori* evidence that white landowners had exchanged profit for racial control.

A comparison of crop yields per acre on owner-operated and tenant farms in Tennessee supports this fundamental premise (see Table 5.5). Overall, yields on owner-operated farms tended to exceed those on rented units, yet the differences in absolute terms were not large. With regard to corn, the superiority of owners over tenants averaged less than two bushels per acre in each region; in Middle Tennessee the yields of owners and

extensive writing on share arrangements and typically uses the latter term to refer to both. In summarizing his analysis I have adopted the same usage, but it should be made clear that his reasoning applies equally to either category of share arrangement.

20. Reid, "Sharecropping as an Understandable Market Response," pp. 126, 129; idem, "Sharecropping in History and Theory," pp. 432–39.

Table 5.5 Crop yields by tenure and race of operator, sample Tennessee
counties, 1879

	East	Middle	West
Median yield per acre			
		Corn (bushels)	
Owner-operators	15.0	20.8	16.7
Fixed-rent tenants	13.3	20.0	15.9
Share tenants/croppers	13.3	20.8	15.0
All white operators	15.0	21.3	17.5
All black operators	15.0	19.4	15.0
		Cotton (bales)	
Owner-operators	NA	0.36	0.44
Fixed-rent tenants	NA	0.38	0.42
Share tenants/croppers	NA	0.40	0.43
All white operators	NA	0.38	0.45
All black operators	NA	0.33	0.41
Adjusted yield index[a]			
		Corn (bushels)	
Owner-operators	12.3	11.8	10.8
Fixed-rent tenants	10.1	10.3	14.9
Share tenants/croppers	12.3	12.2	15.2
All white operators	12.8	11.9	13.0
All black operators	11.5	12.0	13.8
		Cotton (bales)	
Owner-operators	NA	0.20	0.29
Fixed-rent tenants	NA	0.19	0.39
Share tenants/croppers	NA	0.24	0.44
All white operators	NA	0.21	0.33
All black operators	NA	0.20	0.38

[a]Equal to [(median absolute yield/soil quality index) multiplied by 10]. For discussion of
soil quality index, see text.
Source: Eight-county sample, see text.

sharecroppers (and/or share tenants) were actually identical. With regard to cotton, owners' advantage over tenants was statistically negligible in West Tennessee (five to ten pounds per acre), and in Middle Tennessee tenants' yields were actually higher than owners'. Without exception, the observed tenure-associated differences for both crops were small enough to be explained largely (if not wholly) by sampling error, and without question they are far too slight to build a convincing case for either tenant or owner superiority.

The strong showing of tenants and sharecroppers is especially impressive given the likelihood that they often farmed lands of inferior quality compared to that of owner-operated units. Surely most large landholders after emancipation imitated the example of the Middle Tennessee planter who determined to cultivate "thirty or forty acres of the *best*" land himself while renting out the remainder.[21] Soil quality affects yields, and differences among the tenure categories could well account for much of the observed differences in output per acre. Because value per acre provides a rough measure of soil quality, one can test this hypothesis by weighting absolute yield figures for each tenure category by the estimated average value of farm per acre, corrected by linear regression for the average proportion of total acreage unimproved.[22] When crop yields are adjusted to reflect dif-

21. R. S. Ewell to Major Campbell Brown, 10 March 1867, box #1, file #13, Brown–Ewell Papers, TSLA, emphasis added. Thomas J. Edwards argued that the share-cropper was "given the best plot of land upon which to make his crops" but provided no evidence to support his claim. Although Robert Higgs has interpreted Edwards's statement to mean that tenant lands were richer than owner-operated lands, Edwards himself was ambiguous in this regard (the implicit comparison that he makes is to "other classes of farmers"). Given that the article in which his statement appeared was devoted to a comparison of various forms of tenure, it seems more likely that he was comparing the fertility of sharecropped lands versus other types of rented plots. Benjamin Hibbard explicitly claimed that rented land was more valuable than owned land but based his conclusion on aggregate county-level data for 1910, which showed that tenancy rates and overall land values were positively correlated. All that this proved, however, was that the tenancy rate was higher in areas where land values were higher. See Edwards, "The Tenant System and Some Changes Since Emancipation," *Annals of the American Academy of Political and Social Science* 49 (1913), p. 41; Hibbard, "Tenancy in the Southern States," *Quarterly Journal of Economics* 27 (1913):490–1; Higgs, "Race, Tenure, and Resource Allocation in Southern Agriculture," pp. 161ff, fn. 14.

22. Following Roger Ransom, I estimate the value of farm per acre using the following basic equation:

$$V_i = P - C(U_i/T_i),$$

where V_i is the value of farm per acre, P is the price of an improved acre, C is the fixed cost of clearing and breaking an unimproved acre, U equals the number

ferences in soil quality, tenure-related differences are usually reduced and in some instances disappear. Most significantly, in West Tennessee, where tenants operated nearly two-thirds of all farm units, weighted yields for both categories of tenants substantially exceeded those for owner-operators.

Furthermore, black operators compared very favorably with whites overall. According to contemporary opinion, landowners who rented their land to blacks were doubly cursed, first because of inherent flaws in tenancy *per se*, and second because of the inherent unreliability of independent black labor. Even before adjustments for differences in soil quality, however, black farmers reaped yields per acre that were identical to those of whites in East Tennessee and only slightly lower than whites' yields elsewhere in the state. Once absolute yields are weighted to reflect differing average land values, the advantage of white operators entirely vanishes in both Middle and West Tennessee. Whether tenants, and blacks in specific, were also attending conscientiously to work "outside the crop" – that is, projects that would enhance the long-run productivity of the soil – is impossible to determine.[23] What is clear, however, is that Tennessee landowners who turned to tenants for labor could expect short-run returns as great as those they themselves could produce.

Tenants' productivity was high for two probable reasons. First, landlords likely supervised their tenants closely – especially those operating on share contracts – and commonly stipulated contractually the intensity of their labor. Sharecroppers who contracted with Robertson County planter George A. Washington, for example, agreed that the "cultivation & management of said crops & work [were] to be completely under the controul

of unimproved acres, and T equals the total number of improved and unimproved acres. To this basic equation I also introduce dummy variables to control for tenure and racial status, as well as locational dummy variables indicating the county of each operator in order to control (albeit imperfectly) for differences in proximity to markets. See Roger L. Ransom, *Conflict and Compromise: The Political Economy of Slavery, Emancipation, and the American Civil War* (Cambridge: Cambridge University Press, 1989), pp. 72–3.

23. There is simply no systematic evidence available with which to establish the existence of tenure- or race-associated differences in the long-run soil conservation and improvement practices of Tennessee farmers. The complaints of contemporaries as well as modern scholars suggest that exploitative practices were not unique to any racial or tenure group, nor did they emerge for the first time after emancipation. On the exploitative practices of antebellum southerners, see Lewis C. Gray, *History of Agriculture in the Southern United States to 1860* (Washington: The Carnegie Institute of Washington, 1933), pp. 910–11.

of the said Washington."[24] Haywood County's James Bond Jr. required that his tenants

> ... cultivate the land in a judicious manner to prevent injuring it[.] six acres in Cotton & four in corn Peas on gaulded Spots[.] fill gullies with Stalks & brush[,] repair the fences by mooving out the old rails where it is necessary & putting one new rail on each pannel without extra compensation. ... Should parties of second part [sharecroppers] neglect to cultivate or gather their crop & go off of the farm & work elsewhere then the party of the 1st part [Bond] will have the right to charge them with the lost time & hire other hands in their place to cultivate or gather the crop.[25]

Clearly, landlords such as Washington and Bond did not intend to allow their tenants to determine unilaterally the intensity of their labor.

Second, landlords limited the size of their tenants' plots. All other things the same, this should have reduced output per unit of labor and required tenants and their families to work their plots more intensively to achieve a marginal standard of living. The result theoretically was an increased output per acre, which on sharecropped or share-rented land meant an increased return to the landlord. Because the landlord's return per acre was constant under the terms of a fixed-rent contract, his incentive to limit the fixed-rent tenant's plot was less than under a share arrangement, but not nonexistent. Any effort that served to increase per-acre yields decreased the risk that the tenant would default on his payment to the landlord. Consequently, the size of the unit to be rented was of great importance to both landlord and tenant regardless of the form of the rental agreement. As Ransom and Sutch observe, "the ultimate outcome of the bargaining process . . . depended upon the market strength of the two parties, as well as the quality of the land and labor."[26]

24. Quoted in Donald L. Winters, "Postbellum Reorganization of Southern Agriculture: The Economics of Sharecropping in Tennessee," *Agricultural History* 62 (1988):13.
25. Haywood County Records, Office of the Register, Trust Deeds, Vol. 2, 1872–75, mf. roll 75, pp. 468–70.
26. Ransom and Sutch, *One Kind of Freedom*, p. 90. The authors apply this judgment to share arrangements only, maintaining that under the fixed-rent system the rent to be charged per acre was the only issue of true importance, and that after it was agreed on "the renter [might] in principle take on as many acres as he wish[ed]." The key phrase in their hypothesis is "in principle." In a local setting characterized by agricultural risk and an abundant supply of labor, landlords almost certainly would have wished to bargain with regard to both rent per acre and size of plot. See idem, p. 339, fn. 63; and also Gavin Wright, "Freedom and the Southern Economy," *Explorations in Economic History* 16 (1979):101.

Table 5.6 Median improved acres per member of farm-operating
households, by tenure and race of operator, sample Tennessee
counties, 1860-1880

	East	Middle	West
	1860		
All operators	10.0	10.0	8.8
Owner-operators	10.0	10.0	9.1
Tenants	7.5	5.6	6.3
	1880		
All operators	6.7	8.6	6.3
Owner-operators	7.5	10.0	10.0
White	8.3	10.8	10.0
Black	2.0	2.8	8.0
All tenants	5.2	5.6	5.0
White	5.4	6.3	6.3
Black	4.3	4.0	5.0
Fixed-rent tenants	7.5	6.7	5.6
White	8.7	7.5	6.5
Black	a	4.0	5.0
Share tenants/sharecroppers	5.0	5.0	5.0
White	5.4	5.8	6.0
Black	4.7	4.0	5.0

aCell contains fewer than ten cases.
Source: Eight-county sample, see text.

   Long before emancipation, Tennessee landlords regularly allotted their
tenants smaller plots of improved acreage relative to household size than
they reserved for themselves. As per-capita improved acreage figures for
1860 show, tenants' plots were significantly smaller than those of owners
in every section of the state *before* the Civil War. After the war, landowners
simply perpetuated a strategy that had long been widespread (see Table
5.6). The only exception across the state in 1880 involved fixed-rent tenants
in East Tennessee, but that form of tenure was rarely entered into by
eastern farmers and encompassed fewer than 2 percent of farm households
in the region. With the exception of East Tennessee, the difference in
improved acreage per capita between fixed and share arrangements was not
large, supporting the supposition that landlords would have desired to limit

the farm size of both categories of tenants. Overall, postbellum per-capita farm acreage on rented farm units compared to owner-operated farms was 30 percent lower in East Tennessee, 44 percent lower in Middle Tennessee, and fully 50 percent lower in West Tennessee.

As Table 5.6 shows, landlords generally allotted black families fewer improved acres per capita than white families. Why this was so is uncertain. The difference simply may have been the product of prejudice, as land-owning whites consciously sought to discriminate against black tenants. On the other hand, the race-associated difference in improved acres per capita may reflect only the weaker market position of black tenants relative to that of white tenants. Because blacks had fewer nonagricultural opportunities than did whites, a strategy applied equally to both races may have affected blacks more severely.[27]

In reality both factors probably contributed to the racial differential, but a third explanation is also possible. Although tenant contracts frequently stipulated that freedmen work "faithfully" and "industriously," the most effective way for the landowners to insure this was to restrict their acreage to the point that they had to work intensively merely to obtain a subsistence. Landlords wanted white tenants to work just as intensively as black tenants, but they were undoubtedly less suspicious of whites' propensity to work without compulsion. If, as late as 1880, white landowners continued to doubt the ability of blacks to work without supervision, they may have systematically allotted black tenants smaller plots than white tenants, not because of their desire to hold blacks in poverty but because of their determination to maximize their own incomes.

Food Production and Self-sufficiency in 1879

Along with the marked increase in tenancy and a precipitous drop in farm size, a sharp decline in food production was one of the most important features of the postbellum transformation of agriculture. Considering the combined rural population of all eleven ex-Confederate states, per capita production of corn (the leading food crop) dropped by 29 percent between 1860 and 1880, and swine per capita (the primary source of meat) fell by 41 percent.[28] Although this decline in foodstuffs characterized every southern state, in explaining the trend historians have focused chiefly on the

27. Gavin Wright, "Comment on Papers by Reid, Ransom and Sutch, and Higgs," *Journal of Economic History* 33 (1973):175.
28. Wright, *Old South, New South*, p. 35.

major cotton-producing states of the Deep South – the area where the decrease was greatest – and have emphasized above all the increasing concentration on cotton production that occurred there.[29] Depending on the interpretation, farmers either increased their emphasis on the white fiber willingly, as they gambled on high cotton prices to free them from debt, or reluctantly, as the cotton crop became the only basis on which to obtain credit from local merchants.[30] Neither interpretation, of course, can explain the decline of food production in areas that did not produce cotton. For that reason alone, it is important to understand both how and why a major food-producing state of the Upper South was affected.

Statewide figures alone seem to indicate that the disruptive effect was minimal, indeed almost nonexistent. Analysis of aggregate crop and livestock data for Tennessee's ninety-four counties in 1879 yields an index of self-sufficiency in total foodstuffs equal to 1.28, down only 0.02 from 1859.[31] The state's farmers, in other words, produced a total quantity of meat and grain that exceeded human and animal requirements by 28 percent. Despite the extensive physical destruction of agricultural wealth sustained during the war, despite the revolution in the state's labor system that accompanied it and the dramatic decline in scale of farm operations that followed, the state remained lavishly self-sufficient in food production at a level relative to state needs that was almost literally unchanged from the late antebellum period. In sum, patterns of food production appear to have been literally undisturbed by the cataclysm of war and emancipation.

In actuality, nothing could have been further from the truth. Overall, Tennessee farmers did continue to produce an abundance of foodstuffs over and above state demands, but the impression of pronounced continuity that this fosters is wholly illusory, as an examination of household level data from the eight sample counties makes evident. Admittedly, there were a

29. The corresponding figures for the five Deep South states alone (SC, GA, AL, MS, LA) are 40 percent and 46 percent. See ibid., p. 35.

30. For the former interpretation, see Gavin Wright and Howard Kunreuther, "Cotton, Corn and Risk in the Nineteenth Century," *Journal of Economic History* 35 (1975):526–51. For the latter, see Roger Ransom and Richard Sutch, "The 'Lock-in' Mechanism and Overproduction of Cotton in the Postbellum South," *Agricultural History* 49 (1975):405–25.

31. Ten new counties were created between 1860 and 1880, increasing the total for the state from eighty-four to ninety-four. As before, the index of self-sufficiency is equal to the ratio of total food production (expressed in corn-equivalent units) to the estimated amount of total foodstuffs necessary to sustain both livestock and humans. For a discussion of the method, including changes between antebellum and postbellum measurements, see Appendix B.

Table 5.7 Indexes of self-sufficiency in foodstuffs, sample
Tennessee counties, 1859-1879

	East	Middle	West
Entire population[a]			
1859	1.37	1.38	1.14
1879	1.19	1.54	0.99
Farm operators only			
1859	1.80	1.51	1.24
1879 All operators	1.35	1.98	1.35
Whites only	1.38	2.03	1.65
Blacks only	0.95	1.38	0.98

[a]Includes entire farm and nonfarm population.
Source: Eight-county sample, see text.

few basic constants that transcended the divide between Old South and
New and continued to unify the state's three grand divisions. At the *community* level (including both farm and nonfarm population), self-sufficiency
in basic foodstuffs still characterized the state from the Appalachians to the
Mississippi, although among the southwestern cotton counties there was
evidently not an ear of corn or pound of bacon to spare.[32] (See Table 5.7,
figures for "entire population." The tiny apparent deficit in West Tennessee is sufficiently small to be explained wholly by sampling error.) Also

32. It should be noted at the outset that, in terms of their overall indexes of self-
sufficiency, the sample counties did not approximate their larger regions as closely
in 1879 as in 1859. Aggregate county-level data for all of the state's counties yield
indexes of self-sufficiency in total foodstuffs for 1879 of 1.15 in East Tennessee,
1.40 in Middle Tennessee, and 1.21 in West Tennessee. In both East and Middle
Tennessee the discrepancy was still small – 0.04 and 0.14 respectively – and the
changes experienced by the sample counties over the twenty-year period were
similar both in direction and magnitude to those observed for the larger regions.
The sample western counties, however, diverged significantly from their larger
region. Not only was the index of self-sufficiency for Fayette and Haywood counties 0.22 lower than for West Tennessee overall, but the direction of change between 1860 and 1880 was the reverse of that observed for the region as a whole:
The index for the region as a whole rose slightly whereas that for the sample
counties fell sharply. Evidently the experience of the major cotton-producing counties of southwestern Tennessee was markedly different from that of the mixed-
farming areas farther north.

as before the war, farm *operators* in every section of the state not only remained comfortably self-sufficient in foodstuffs but also continued to produce surpluses far above that required merely for safety-first behavior.[33] Farm operators in both East and West Tennessee produced one-third more foodstuffs than required for on-farm consumption, and Middle Tennessee operators actually produced double their household needs. Except for West Tennesseans, farmers continued to export the bulk of their surplus to external rather than local markets.[34]

Such elements of continuity aside, the complexity and diversity of change after 1860 are far more striking. The sample microlevel data suggest that patterns of food production underwent extensive alterations between 1860 and 1880, that these alterations were complicated and occasionally surprising, and that they varied substantially across the state. At the community level, a comparison with 1860 figures shows that the ratio of food produced to food required for subsistence declined sharply in both East and West Tennessee but rose almost as markedly in the state's Central Basin. Although, at first glance, the decline of food production in the mixed-farming area of East Tennessee – the "Switzerland of America" – is more than a little puzzling, the dwindling supply of foodstuffs in West Tennessee accords perfectly with a chorus of contemporary complaints regarding an unprecedented obsession with cotton among western farmers. For example, at the conclusion of the first peacetime harvest, a Hardeman County newspaper denounced the growing infatuation with "cotton, cotton, cotton . . . the blood sucker of all that is made in this section." A few months later, the perceived trend apparently undiminished, the editor lamented that it would soon be "almost an impossibility to purchase anything produced in this section for the sustenance of life."[35]

Such complaints continued unabated into the next decade. In the mid-1870s the state's commissioner of agriculture, James B. Killebrew, compared West Tennessee farmers unfavorably with those in other sections of the state. The typical farmer in East or Middle Tennessee, according to Killebrew, "aims at making a supply" of foodstuffs, whereas the average farmer of southwestern Tennessee "cares for nothing so much as to see his

33. For a discussion of safety-first behavior, see Wright and Kunreuther, "Cotton, Corn and Risk in the Nineteenth Century."
34. Provided that all food-deficient farm and nonfarm households purchased necessary foodstuffs locally, local demand still could have absorbed only about 33 percent of the surplus in East Tennessee and 29 percent in Middle Tennessee.
35. *Bolivar Bulletin*, 24 November 1865, 17 January 1866.

cotton fields flourishing. . . . He will stake his all upon his prospects for cotton; chickens, eggs, butter, corn, wheat, hay, meat – all these are little things, and cotton will buy them."[36]

Significantly, however, an examination of self-sufficiency indexes among *farm operators only* – as opposed to the entire farm and nonfarm populations – yields markedly different insight into regional distinctions and casts considerable doubt on the validity of such jeremiads with regard to West Tennessee. (See Table 5.7, figures for "farm operators only.") A comparison with 1860 figures shows that operators had increased their output of foodstuffs in not one but two of the state's three major regions. Unexpectedly, the region where food output remained lower was not West Tennessee – where nine out of ten farmers devoted considerable attention to a nonfood crop – but East Tennessee, a region where cotton and other such "blood sucking" staples were virtually nonexistent. Indeed, an examination of race-specific figures for 1879 indicates that, far from reducing their emphasis on food production, *white* farm operators in West Tennessee had dramatically increased their output of foodstuffs since before the war, presumably to accommodate the growing number of households (predominantly black) without direct ties to the land.[37] The overall result was that, by 1880, the cotton growers of the Tennessee Black Belt (both white and black) and the mountain farmers of Appalachian East Tennessee were statistically indistinguishable in their collective output of grain and meat.

Although in the aggregate farm operators in each region of the state still produced ample quantities of food for their families, the proportion of individual farmers deficient in basic foodstuffs was sizable. Table 5.8 presents detailed self-sufficiency rates for the crop year 1879, broken down by race and tenure categories. The table contains two estimates for each subcategory: one that makes no allowance for rental payments by tenants in the form of grain, and another (in parentheses) derived by reducing the grain output of fixed-rent tenants by one-third and that of share tenants and sharecroppers by one-half.[38]

---

36. Killebrew, *Introduction to the Resources of Tennessee*, pp. 357, 359.
37. Undoubtedly, some portion of the apparent surplus among white operators in West Tennessee was allocated to hired (black) farm laborers who worked for wages plus board. Although it is possible to use wage data from the census to estimate roughly the number of full-time adult wage hands (or their equivalent) employed per white-operated farm, I have chosen not to adjust the indexes for white operators in Table 5.7 on the grounds that in the aggregate white farmers received additional foodstuffs in the form of rent that would have offset, or very nearly offset, the quantity of foodstuffs distributed to farm employees.
38. According to Ransom and Sutch, sharecroppers generally were not responsible for

Table 5.8 Proportion of farm operators self-sufficient in total foodstuffs, sample Tennessee counties, 1879 (percentages)

	East	Middle	West
All operators	53.8 (46.4)	76.4 (70.3)	55.7 (41.0)
White	55.8 (48.8)	79.4 (74.4)	73.7 (64.5)
Black	37.4 (26.7)	53.2 (38.9)	40.6 (21.2)
Owner-operators	53.8 (NA)	79.2 (NA)	73.6 (NA)
White	56.3 (NA)	81.6 (NA)	75.9 (NA)
Black	28.6 (NA)	41.8 (NA)	59.2 (NA)
Fixed-rent tenants	49.5 (33.7)	65.2 (52.8)	49.9 (29.3)
White	49.5 (33.7)	68.9 (56.3)	67.9 (42.3)
Black	*a* *a*	50.0 (38.3)	42.2 (23.6)
Share tenants/croppers	54.4 (22.5)	71.2 (44.7)	40.2 (14.0)
White	54.9 (22.1)	74.2 (47.9)	72.4 (32.1)
Black	52.4 (24.2)	63.8 (36.7)	35.5 (11.4)

*a*Cell contains fewer than ten cases.
Source: Eight-county sample, see text.
Note: Figures in parentheses are adjusted rates after grain production is reduced to reflect rent payments by tenants. See text.

Rent-adjusted rates of self-sufficiency fell in every section of the state during the twenty years after 1860. The decline was least pronounced in Middle Tennessee, the "garden" of the state, where the adjusted rate of self-sufficiency dropped by fewer than 5 percentage points. In the sample counties of the Central Basin, seven-tenths of farm operators still produced more than enough food for household needs. Elsewhere the change was far more substantial, however; between 1860 and 1880 rent-adjusted rates fell by nearly 25 percentage points in West Tennessee and by almost 30 percentage points in East Tennessee. Although farm operators in both regions were self-sufficient in the aggregate, a *majority* of individual farm families were not, helping to explain the frequent cries for agricultural reform noted earlier. Before the Civil War, farmers in East and Middle Tennessee had

feeding work stock, which was typically provided by the landlord. I have thus reduced accordingly the grain requirements of all farmers operating on share arrangements, although to the extent that these individuals were share tenants (who owned and were responsible for feeding their own work stock) rather than share-croppers, the rates of self-sufficiency for "Share Tenants/Croppers" in Table 5.8 will be slightly overstated. See Ransom and Sutch, *One Kind of Freedom*, pp. 251–3.

Table 5.9 Proportion of **white** farm operators self-sufficient in total food-stuffs, sample Tennessee counties, 1859-1879 (percentages)

	East		Middle		West	
All white operators						
1859	79.7	(76.6)	78.2	(74.7)	68.9	(65.1)
1879	55.8	(48.8)	79.4	(74.4)	73.7	(64.5)
White owner-operators						
1859	80.9	(NA)	82.0	(NA)	69.7	(NA)
1879	56.3	(NA)	79.2	(NA)	75.9	(NA)
White tenants[a]						
1859	73.7	(55.3)	63.1	(45.8)	63.5	(36.3)
1879	54.4	(23.2)	72.2	(51.1)	69.0	(39.7)

[a]Weighted average of fixed-rent tenants and share tenants/sharecroppers.
Source: Eight-county sample, see text.
Note: Figures in parentheses are adjusted rates after grain production is reduced to reflect rent payments by tenants. See text.

been virtually identical in their propensity for self-sufficiency; farmers in West Tennessee had trailed behind somewhat but had still compared favorably with northern farmers in such leading food-producing states as Indiana or Minnesota. By 1880, in contrast, the proportion of self-sufficient operators was 24 to 29 percentage points higher in Middle Tennessee than elsewhere, and farmers in East and West Tennessee – regions that could not have differed more radically in racial composition, extent of tenancy, or emphasis upon cotton – were highly similar in their far lower rates of self-sufficiency.

Although the magnitude of the decline in self-sufficiency was comparable in the two disparate regions, the nature and apparent cause of the decline differed considerably. Significantly, even though scholars have often linked the postbellum decrease in food production with the concomitant growth in tenancy, it would be wrong to attribute declining self-sufficiency in either East or West Tennessee primarily to an increase in tenancy *per se.* In East Tennessee the drop-off chiefly reflected a pervasive erosion of self-sufficiency among whites that afflicted *both* owner-operators and tenants. Although the decline in rent-adjusted rates among the latter was a staggering 34.5 percentage points, the percentage-point drop among the former – 27.1 – was very nearly as large (see Table 5.9). In West Tennessee, on the other hand (and Middle Tennessee as well), falling rates of self-sufficiency were almost entirely due to the performance of black households. Indeed, in the major cotton-producing counties of southwestern Tennessee, the rate of self-sufficiency among white operators – both owners and tenants – actually increased after the Civil War. In contrast, black operators of all tenure categories exhibited self-sufficiency rates two-thirds lower than those of antebellum whites (see Table 5.8).

The lower rate of self-sufficiency among blacks relative to whites was not limited to the Cotton Belt only. After deductions for rent, whites in East and Middle Tennessee were also about twice as likely as blacks to be self-sufficient in foodstuffs. In all three regions of the state, part of the overall dissimilarity simply reflected the fact that the freedmen were disproportionately tenants, and tenants (of both races) were less likely to be self-sufficient than were owner-operators. There was more to the disparity between the races, however, for within each tenure category self-sufficiency continued to be substantially lower among blacks than among whites. (The share category in East Tennessee was an exception that proved the rule.)

A number of factors potentially could explain the vastly lower incidence of self-sufficiency among blacks relative to whites. Contemporaries often pointed to the former slaves' "overproduction" of cotton. A West Tennessee newspaper, for example, complained that the freedmen were "so wedded . . . to the raising of cotton that they will not contract with a farmer who proposes to produce the necessary articles for home consumption."[39] Although the cause was likely not so straightforward as the editor maintained, it is true that as late as 1879 blacks in the sample western counties did devote a significantly larger proportion of their improved acreage to cotton than did whites.[40] This greater emphasis on cotton may provide a partial explanation for food deficiencies among West Tennessee freedmen, but it clearly cannot adequately explain insufficient food production among blacks elsewhere in the state. In the sample counties of East and Middle Tennessee, blacks who either desired or were forced to increase their emphasis on a cash crop primarily redoubled their efforts to produce foodstuffs, not cotton. Consequently, in these regions, and perhaps in West Tennessee as well, some other factor or factors must explain race-associated differences in food production.

One possible explanation was the smaller number of improved acres per

39. *Bolivar Bulletin*, 17 January 1866.
40. Overall, in West Tennessee blacks planted 54.4 percent of their improved acreage in cotton, whites 39.0 percent.

capita typically allotted to black families (see Table 5.6).[41] Other things the same, black farmers should have had more difficulty in attaining self-sufficiency in grain than did whites, who had relatively greater tillable acreage. Another obvious factor, which reflected blacks' relative poverty, was their smaller holdings of beef cattle and hogs. Across the entire eight-county sample, blacks in 1880 owned fewer than half as many hogs and cattle per capita as did whites, a not particularly startling finding considering that only fifteen years earlier they had owned no livestock at all. It should not be surprising, then, if meat production was considerably lower among black tenants and croppers.

Table 5.10 presents standardized regression coefficients[42] from a multiple regression equation constructed to estimate the relative importance of these factors in explaining variations in food production on Tennessee farms. The dependent variable in each equation is net per-capita food output, defined as the per-capita quantity (in corn-equivalent bushels) of grain and meat available for human consumption after allowances for seed and feed for livestock. Independent variables in each equation include the number of improved acres per household member, the per-capita value of livestock, the proportion of improved acreage planted in cotton, and dummy (dichotomous) variables indicating the tenure and race of each operator.

Contrary to expectations, the results of the regression analysis reveal that in West Tennessee the independent influence of the crop mix on food output was negligible. After controlling for variations in farm size and livestock holdings, the proportion of improved acreage planted in cotton

41. Several farm management studies conducted early in the century pointed to size of farm as a critical factor in determining farm output and income. See E. A. Boeger and E. A. Goldenweiser, "A Study of the Tenant Systems of Farming in the Yazoo–Mississippi Delta," *United States Department of Agriculture Bulletin*, no. 337 (Washington, DC: Government Printing Office, 1916), pp. 9–18; H. M. Dixon and H. W. Hawthorne, "An Economic Study of Farming in Sumter County, Georgia," *USDA Bulletin*, no. 492 (Washington, DC: Government Printing Office, 1917), pp. 35–64; and E. S. Haskell, "A Farm-Management Survey in Brooks County, Georgia," *USDA Bulletin*, no. 648 (Washington, DC: Government Printing Office, 1918), pp. 18–20.

42. Because not all the independent variables in the equation are measured in the same units, regression coefficients generated by the equation are not strictly comparable. To evaluate the relative explanatory importance of each independent variable, it is necessary to examine standardized regression coefficients, or beta weights, which are essentially dimensionless coefficients. The standardized coefficient equals the regular B-coefficient multiplied by the ratio of the standard deviation of the independent variable to the standard deviation of the dependent variable.

Table 5.10 Determinants of net per-capita food production among farm operators, sample Tennessee counties, 1879

	East	Middle	West
	Standardized regression coefficients (t-values in parentheses)		
Improved acres per capita	0.400[a] (15.0)	0.681[a] (29.9)	0.468[a] (22.8)
Livestock value per capita	0.464[a] (17.5)	0.013 (0.6)	0.423[a] (20.0)
Percentage of improved acres planted in cotton	0.005 (0.3)	0.008 (0.5)	0.031 (1.7)
Dummy variable indicating black operator	-0.053[a] (-2.9)	0.034 (1.8)	0.090[a] (4.0)
Dummy variable indicating fixed-rent tenant	0.009 (0.5)	0.022 (1.2)	-0.033 (-1.4)
Dummy variable indicating share tenant or sharecropper	0.057[a] (3.1)	0.017 (0.9)	-0.012 (-0.5)
Coefficient of Determination (R^2)	0.62	0.46	0.67

[a]Significant at 0.05 level of confidence.
Source: Eight-county sample, see text.

contributes little to explained variation. In the Black Belt counties of southwestern Tennessee, blacks produced less food per capita than did whites primarily because they owned less livestock and farmed smaller plots, not because of their tendency to devote more of their acreage to cotton.[43] Indeed, because the standardized coefficient for the dummy variable denoting black operators is both positive and statistically significant, it appears that

43. Regression analysis of farm-specific crop mix data, controlling for variations in farm size and race and tenure status, further supports this conclusion. The equation generated indicates that, when variations in farm size and tenure are held constant, blacks were actually less likely than whites to emphasize cotton over food crops. The equation is:

$$\text{Crop Mix} = 0.457 - 0.001A - 0.081B + 0.099F + 0.099S, R^2 = 0.11$$
$$\phantom{\text{Crop Mix} = } (23.8) \quad (2.1) \quad (4.6) \quad (5.3) \quad (4.6)$$

where crop mix equals the proportion of improved acres planted in cotton, A equals the number of improved acres per household member, and B, F, and S are dummy variables indicating blacks, fixed-rent tenants, and share tenants (and possibly sharecroppers), respectively (t-statistics in parentheses).

had the freedmen in West Tennessee worked farms comparable in size to those of whites and owned similar amounts of livestock, they actually would have produced more food per capita than whites did.

Variations in crop mix notwithstanding, the primary factors influencing food production appear to have been pretty much the same all across the state. For the East Tennessee data, improved acreage and livestock value per capita dwarf all other variables in explanatory power; for Middle Tennessee the acreage variable alone towers above all others. The coefficient for the dummy variable distinguishing blacks is insignificant for Middle Tennessee; for East Tennessee it is both significant and negative, yet the independent effect of race was evidently of a far smaller magnitude than that of either farm size or livestock holdings. Significantly, in none of the regions was tenure status *per se* negatively correlated with food production. For both Middle and West Tennessee the dummy variables indicating tenants have statistically insignificant coefficients; for East Tennessee the coefficient for the share dummy is significant but positive, indicating that, other things the same, share tenants and sharecroppers produced more food per capita than other farmers did.

Variations in per-capita improved acreage and livestock holdings may explain output variations at a single point in time, 1879, but this does not prove that *changes* in those variables were responsible for the proportional decrease in self-sufficient households after 1860. It is difficult to test such a hypothesis rigorously, but indirect evidence suggests that it has some validity. First, regression analysis of antebellum data reveals that in both East and Middle Tennessee improved acreage per household member was also the single most influential determinant of food production in 1859, followed in importance by livestock holdings per capita. In West Tennessee, where farmers were tempted to forgo self-sufficiency in order to concentrate on cotton, the crop mix was a more crucial factor than commitment to livestock, yet both variables were far less important than improved acreage per capita in explaining variations in food output.[44]

44. These conclusions are based on the following regression equations, where F equals net per-capita food production (in corn-equivalent bushels), A represents improved acreage per capita, C is the number of pounds of cotton grown per bushel of corn, L equals value of livestock per capita, and T is a dummy variable distinguishing tenants from owner-operators (t-statistics in parentheses).

East Tennessee:
$$F = 11.3 + 3.5A + 0.3L + 0.1C - 1.8T, R^2 = 0.59$$
$$(3.4) \quad (19.0) \quad (12.3) \quad (0.03) \quad (-0.2)$$

Second, it is probably not coincidental that among white operators the rate of self-sufficiency fell noticeably only in East Tennessee, the one region in the state where both improved acreage and livestock per capita among whites also dropped sharply. In both Middle and West Tennessee, where self-sufficiency rates among whites either remained constant or increased between 1860 and 1880, major departures from antebellum patterns in per-capita farm size and livestock holdings were restricted largely to black-operated farms.[45]

Gross Farm Income in 1879

As this analysis has shown, civil war and emancipation initiated major changes in agricultural organization that transformed each of Tennessee's major divisions in varying degrees. The period between 1860 and 1880 witnessed a sharp decline in the scale of farm production, a concomitant rise in the importance of agricultural tenancy, and a significant increase in the proportion of farm households unable to feed themselves. Not surprisingly, patterns of farm income also underwent extensive change.

To begin with, the average value of production on Tennessee farms fell drastically (see Figure 5.2). The proportional decline was greatest in West Tennessee, where the value of farm output (adjusted for changes in farm prices) fell by two-thirds.[46] It was severe in all sections, however; the value

Middle Tennessee:

$$F = 23.3 + 3.3A + 0.2L - 2.3C + 16.1T, \; R^2 = 0.28$$
$$\;\;\;\;\;\; (6.7) \;\; (15.2) \;\; (8.0) \;\; (-2.2) \;\;\;\;\; (2.9)$$

West Tennessee:

$$F = 30.8 \;\; + 2.4A + 0.1L - 0.7C + 9.4T, \; R^2 = 0.46$$
$$\;\;\;\;\;\; (12.0) \;\; (13.5) \;\; (4.3) \;\; (-8.4) \;\;\;\;\; (2.2)$$

45. The per-capita number of swine on white-operated farms fell from 3.4 to 2.0 in East Tennessee, decreased from 3.6 to 3.1 in Middle Tennessee, and actually increased from 2.9 to 4.3 in West Tennessee. The dramatic change in West Tennessee was a statistical by-product of emancipation. Before the war, slaves had been included in the households of their owners and thus had figured into the computation of per-capita figures. After emancipation, the freedmen left not only their former masters but their supply of meat as well. For a similar argument that declining farm size was probably the single greatest factor behind declining self-sufficiency, see Wright, *Old South, New South*, pp. 111–12.

46. To control for changing commodity prices, the 1879 production values represented in Figures 5.2 and 5.3 were constructed using 1859 wholesale price data.

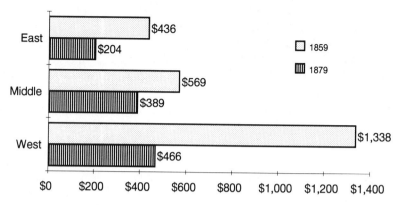

Figure 5.2   Median total value of production per farm in constant dollars, sample Tennessee counties, 1859–79

of output per farm unit dropped by one-third in Middle Tennessee and by one-half in the easternmost part of the state. In all three regions the decline was almost entirely a result of the dramatic decrease in farm size that characterized the period, as the relatively constant values of output per improved acre make clear (see Figure 5.3). Because the overall decline in farm size varied among the regions and was greatest in the southwestern cotton counties, one of the effects of declining scale was to lessen substantially the differences among the regions. For example, the absolute dollar difference in total production between eastern and western farms declined by more than two-thirds over the twenty-year period, from $902 to $262.

As before the war, income from agricultural production was distributed unevenly among the farm populations of each region. Working from the values of farm production represented in Figure 5.2, Table 5.11 presents per-capita income figures for various farm categories. As for the 1859 estimates presented in Chapter 1, the value of total production for 1879 has been reduced to account for the total value of crops fed to livestock, retained for seed, and, in the case of tenants, paid to the landlord as rent. Additionally, new information provided for the first time by the census in 1880 allows further adjustments for expenditures for hired labor and fertilizer. These figures, it should go without saying, are rough approximations only. More precise estimates would also account for capital gains on land and improvements, the implicit rental value of housing, the natural increase of livestock inventories, and the depreciation of farm tools, wagons, and machinery. Finally, it is important to emphasize that the income figures measure only the approximate value of income earned *directly* from the unit

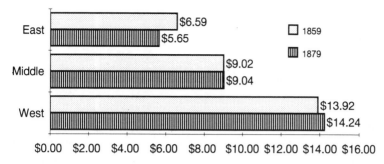

Figure 5.3   Median total value of production per improved acre in constant dollars, sample Tennessee counties, 1859–79

of production; they do not take into consideration income that farmowners may have earned by renting a portion of their lands to other operators. As a result, income differences within each region – between owners and tenants or black and white operators, for example – were undoubtedly larger than the figures on farm production per capita would suggest. The same would be true of differences among the regions because tenancy was relatively much more extensive in West Tennessee than elsewhere.

Despite the unparalleled changes that had transformed farming in all three regions since 1860, the overall differences in income between the regions continued to be substantial. Although by 1880 the antebellum size advantage of West Tennessee farms had entirely disappeared, farmers in the section continued to hold a decided edge in per-capita income over those elsewhere in the state. The reason is straightforward: West Tennesseans concentrated far more heavily on cotton, and cotton cultivation, of all the uses to which agricultural land could be put, was by far the most valuable. For the state as a whole, the average value per acre planted in cotton was nearly triple that for an acre in corn.[47] As a result, even without allowing for rent income – which must have been proportionally much greater in West Tennessee than elsewhere – per-capita income in the sample cotton counties was two-thirds greater than in Middle Tennessee and four times greater than in East Tennessee.

To place the figures in a meaningful context, it is again useful to employ

47. The average per-acre yield of cotton in Tennessee in 1879 was worth $26.80; the corresponding figure for an acre of corn was $9.25. See *Biennial Report of the Commissioner of Agriculture, 1889–1890* (Nashville, 1891), pp. 49, 51; Wright, *Old South, New South*, p. 34.

Table 5.11 Median farm income per capita in farm-operating households, sample Tennessee counties, 1879

	East	Middle	West
All operators	$13	$34	$54
Owner-operators	15	42	85
White	15	44	83
Black	7	14	99
Fixed-rent tenants	7	19	54
White	7	20	60
Black	a	13	51
Share tenants/croppers[b]	9	20	35
White	9	22	37
Black	10	16	35
All white operators	13	38	73
All black operators	7	16	44
Market-oriented[c]	48	56	57
Subsistence-oriented[d]	9	13	4

[a]Cell contains fewer than ten cases.   [b]Listed ambiguously on the agricultural schedule as renting "for shares of products," but may have included both share tenants and sharecroppers.   [c]Able to sell one-half or more (by value) of total farm production off the farm after meeting subsistence needs.   [d]Unable to sell as much as one-half (by value) of total farm production off the farm after meeting subsistence needs.
Source: Eight-county sample, see text.

the consumption level of antebellum slaves as a frame of reference. As mentioned in Chapter 1, the typical Tennessee slave in 1859 consumed $28 to $30 worth of food and nonfood items (exclusive of the value of housing). Taking into consideration changes in the cost of living over the next twenty years, Tennessee farm families must have earned, at a very minimum, at least $21 per capita in order to approximate the material standard of living of antebellum slaves.[48]

Given this benchmark for comparison, it is abundantly clear that by 1880 the mountain farmers of East Tennessee were facing an agricultural crisis of critical proportions. The pressure of a rapidly growing population against

48. The cost-of-living adjustment takes into consideration changes in wholesale prices of corn and pork in major Tennessee markets but otherwise rests on the estimates of Ransom and Sutch concerning the cost of living of black agricultural workers in the Cotton South. See *One Kind of Freedom*, p. 218.

a severe land constraint had not only plunged the size of farms downward but had also forced increasingly marginal lands into cultivation as prospective farmers were pushed farther up the mountainsides and deeper into the hollows.[49] The result, for a majority of farmers at least, was that farm operations alone could no longer provide adequately the necessities of life. Many did not even come close. Tenants of both colors, for example, earned incomes from the land that were worth less than half as much as the consumption of antebellum slaves. Even farmowners, particularly those who earned no rental income, commonly required off-farm employment to supplement their incomes. Certainly some East Tennessee operators would have been able to augment their earnings by working occasionally by the day on neighboring farms. Yet, considering the large population of non-operators already searching for work, it is highly doubtful that such opportunities would have been lucrative.

In addition to differences among the regions, differences within each region are also impressive. As one would expect, per-capita income was uniformly greater for owner-operators than for tenants or croppers, and, with one exception, uniformly greater for whites than for blacks within each tenure category. Because the freedmen were disproportionately tenants, the overall disparity between the races was even larger; across the sample counties as a whole, white incomes from crop and livestock production exceeded those of blacks by a ratio of nearly 2:1.[50]

The figures in Table 5.11 prompt two additional observations. First, an interregional comparison suggests that the freedmen who migrated within Tennessee from mixed-farming regions to the Cotton Belt made economically wise choices. Not only was the income disparity between the races smallest proportionally in the Black Belt, but the absolute level of farm income enjoyed by black operators was also far higher there than anywhere else in the state. This was primarily due to their ability to plant cotton, an option that geography and climate denied freedmen in most other parts of Tennessee. In some ways this was a dubious opportunity, of course, in that it exposed the freedmen of West Tennessee to the vagaries of the postbellum cotton economy. Nevertheless, it is clear that in the short run at

49. I elaborate on this point in "Wealth and Income: The Preindustrial Structure of East Tennessee in 1860," *Appalachian Journal* (forthcoming). See also Altina Waller, *Feud: Hatfields, McCoys, and Social Change in Appalachia, 1860–1900* (Chapel Hill: University of North Carolina Press, 1988).

50. It must be emphasized that these income figures do not incorporate landlords' income from rent, and thus they underestimate the racial differential in total income per capita.

least the freedom to plant cotton gave West Tennessee blacks a decided advantage. Most important, it helped to put bread on their tables. Although blacks in West Tennessee were less likely to achieve self-sufficiency than in other sections, nearly nine-tenths produced enough cotton to pay for any additional food that their families required, even with exorbitant credit prices more than 50 percent above wholesale levels.[51] The vast majority of freedmen in East and Middle Tennessee, on the other hand, had no choice but to find additional sources of income from off their farms.

Finally, as was also true of antebellum income patterns, the most conspicuous differences were not between regions but within each region between those who produced extensively for the market and those who, for reasons of choice or necessity, produced primarily for family consumption. Table 5.11 compares the median net per-capita income of market-oriented households, those that sold one-half or more of their output, with that of subsistence-oriented households, families that sold less than one-half of all that they produced. In each region the income disparity between the two groups can only be described as astounding. Market-oriented households earned incomes more than five times greater than subsistence-oriented households in East Tennessee, more than four times greater in Middle Tennessee, and more than fourteen times greater in West Tennessee. The income from crop and livestock production among the subsistence-oriented was so desperately low that they must have supplemented their earnings substantially with off-farm income; as a consequence, the inordinate disparities given in Table 5.11 exaggerate to some unknown extent the true income differential between the two groups. Nevertheless, the evidence is compelling that poverty, rather than independence, was the real reward of those truly isolated from the market. That southern farmers in significant numbers should have aspired to such a state defies the imagination.

Conclusion

To a surprising extent all three of Tennessee's grand divisions shared in the most prominent trends associated with the postemancipation reorgan-

---

51. Although there is little good evidence concerning retail credit charges in Tennessee, detailed state records from Georgia show that during the 1880s Georgia merchants on the average charged a credit price for corn that was 53 percent above the wholesale farm-gate selling price. Assuming a similar price structure in Tennessee for both corn and pork, and after making the appropriate reductions to reflect the costs of rent and of ginning and baling, 88 percent of West Tennessee freedmen still earned enough from their cotton to buy all the food that they needed without carrying over any debt into the subsequent crop year. For the Georgia price data, see Ransom and Sutch, *One Kind of Freedom*, p. 129.

ization of agriculture. In varying degrees all three sections experienced an increase in the number of farm units, due partly to an increase in tenancy but also to a sharp rise in the number of white-owned farms. Considering that improved farm acreage grew modestly, a related development in each region was a sharp decline in the average size of farm units, a factor that apparently contributed instrumentally to a widespread decline in household food self-sufficiency, a trend that was particularly pronounced in East and West Tennessee. Although scholars have frequently attributed declining food output to a change in commercial orientation among individual producers, in Tennessee larger impersonal developments seem to have been more critical. Dwindling supplies of land and livestock – the first due to the pressures of population growth against a relatively rigid land constraint, the second primarily the by-product of military destruction – were the primary deterrents to household self-sufficiency. As before the war, the most striking differences among the regions continued to be in the level of income that farm operations afforded. Despite the greater upheaval in the western cotton counties due to the demise of slavery, cotton cultivation continued to reward western farmers, both black and white, at a level that far exceeded that of farmers anywhere else in the state.

Conclusion:
One South or Many?
Implications for the
Nineteenth-century South

It is customary for authors of case studies to conclude by establishing the representativeness of their subjects and the broader significance of their findings. At that point, as I have long imagined but now understand experientially, there is a powerful impulse to overstate the results of one's work by generalizing broadly and uncritically to some larger historical universe – the nineteenth-century South, for example. Implicitly, scholars who yield to such temptation assume that the larger universe of interest was reasonably homogeneous and that in describing a portion they have also described the whole.

Although it is impossible to know in many cases, I suspect that the assumption of homogeneity – with regard to historical analysis, anyway – is more often than not inappropriate. In this particular instance I know it to be so, for it denies the very diversity that this book has sought to investigate. In closing, therefore, it is imperative to review briefly the conclusions of the preceding chapters and to suggest what implications they hold, if any, for the South as a whole and, by extension, for the future study of southern history.

If anything, this analysis of Tennessee in the Civil War era has shown the futility of blanket comparisons of plantation and nonplantation districts. On the eve of the Civil War, Tennessee's white agricultural population did vary considerably from region to region, most noticeably in dependence on slavery. This disparity in turn contributed to other significant distinctions with regard to work routine, prevalence of large plantations, extent of wage labor, commitment to market production, and level of farm income. The claims of contemporary Tennesseans concerning their state's rich diversity were clearly well founded.

190

At the same time, however, it would be easy to exaggerate such inter-regional variations and misinterpret their significance. Despite glaring dif-ferences, there were also fundamental similarities that unified the state's grand divisions in 1860. The typical scale of farm operations, for example, was roughly comparable from region to region, as was the frequency with which farm households achieved basic self-sufficiency in foodstuffs. Even more important was the striking resemblance among the regions in local socioeconomic structure. Despite great disparity in both the magnitude and composition of wealth, local distributions of wealth were almost identical from section to section, varying negligibly from the Appalachian foothills to the hinterlands of Memphis. Nor did the extent of socioeconomic mo-bility (i.e., patterns of property acquisition and accumulation) differ sub-stantially across the state. Modest but widespread upward mobility, framed within an overall context of impressive socioeconomic stability, character-ized the local white farm populations of both Cotton Belt and mixed-farming regions. Of course, whether these eight sample counties were typical of plantation and nonplantation areas is uncertain, and without ex-tensive comparable study by other scholars, it is ultimately unprovable. Emphatically, however, the antebellum patterns observed in this investi-gation indicate that there are no *a priori* reasons to link the extent of relative inequality or social stratification in a community to that community's extent of reliance on slavery, degree of market integration, or absolute level of wealth. None of these correlations obtained in the disparate areas studied.

Above all else, I hope that this finding will encourage historians to re-examine the element of nostalgia that pervades much of the "new rural history."[1] What I have in mind is not so much the romanticizing of rural life *per se* as the idealization of purportedly precapitalist rural communities.[2] In particular, numerous scholars now argue that communities untouched by the market economy were characterized by patterns of cooperation and other manifestations of virtue that were absent in more economically de-veloped regions. Such claims are incredibly difficult to document and, to my mind, rest on excessively optimistic appraisals of human nature in the absence of corrupting external forces. Even if the comparison is accurate, however – that is to say, even if areas removed from the market economy

---

1. For an early evaluation of this new genre, see Robert P. Swierenga, "The New Rural History: Defining the Parameters," *Great Plains Quarterly* 1 (1981):211–23.

2. On the issue of rural nostalgia, see Steven Hahn and Jonathan Prude, eds., *The Countryside in the Age of Capitalist Transformation: Essays in the Social History of Rural America* (Chapel Hill: University of North Carolina Press, 1985), p. 7.

did exhibit kinder, gentler social relationships – this finding would still tell only part of the story. Despite the recent outpouring of work on rural America, we continue to lack sufficient systematic data concerning the economic costs of market isolation, either to individual farmers or to entire communities. This deficiency is critical, because without adequate information concerning the disadvantages of market estrangement, we risk an unbalanced, even distorted evaluation of the impact of market integration.

For nineteenth-century Tennessee, at least, the economic implications of market isolation are abundantly clear. They are revealed less by a comparison of communities – none of the sample counties in this study was truly isolated from external markets – than by a comparison of subsistence- and market-oriented farmers within the same locale. Simply put, both before and after the Civil War, farmers who were either unwilling or unable to participate in larger networks of exchange were invariably poor, earning incomes that were 70 to 90 percent lower than those of market-oriented farmers in the same region. Such farmers may have supplemented their diet with garden products and wild game, and they may have augmented their income with off-farm employment, yet the basic conclusion still stands: Among farmers who sold less than one-half of their output in the marketplace, the average income generated by crop and livestock production alone was insufficient to sustain a material standard of living appreciably superior to that of the typical plantation slave. If market isolation fostered greater economic independence among certain Tennessee farmers – a thesis not convincingly established, in my opinion – the evidence is incontrovertible that it also bred desperate poverty.

Again, it is uncertain whether the patterns observed in this study were widely duplicated in other parts of the South, yet there is considerable reason to believe that they would have been. Although historians have frequently waxed eloquent about the "lavish" self-sufficiency of the southern plain folk, it is doubtful that there were ever very many "self-sufficient" communities that were not also, by definition, poor.[3] It simply made no sense for farmers without guaranteed markets to produce significantly more than their families could consume. Consequently, it is nearly axiomatic that communities without stable ties to external markets – or, at the very least, extensive local demand for foodstuffs due to large nonfarm populations – were communities on the margin of subsistence. We need not assume that

3. Rodney C. Loehr, "Self-Sufficiency on the Farm," *Agricultural History* 25 (1952): 37–41.

southern whites were profit-driven entrepreneurs to admit that, given the costs of market isolation, most producers would have welcomed opportunities for commercial exchange. The typical smallholder may indeed have valued autonomy and thus sought to escape the dictates of the market. He may also have cherished the financial security of his family and thus sought to avoid the possible abuses of the market. But that a sizable number would have rejected the market *per se* and the potential benefits that market exchange afforded is a hypothesis both incredible and unproven.

In sum, although I applaud scholars' growing appreciation of antebellum southern diversity and their refusal to view the history of the South as the history of the Cotton Belt writ large, this investigation of interregional differences in Tennessee raises the possibility that recent scholarship both oversimplifies distinctions between plantation and nonplantation regions and exaggerates the socioeconomic heterogeneity of the Old South. The point is not that the differences between plantation and small-farming regions were unimportant but that there were striking similarities as well.

This same complexity also militates against sweeping interregional comparisons with regard either to the impact of civil war and emancipation or to subsequent structural features of the postbellum South. War and emancipation affected each of Tennessee's major regions in different ways and to different degrees. Quite obviously, the social and economic disruption generated by emancipation was far greater in West Tennessee than elsewhere, yet the destruction and destitution caused by contending armies seems to have varied inversely with the antebellum importance of slavery. Despite such obvious differences, however, to different degrees all three regions shared in the dominant agricultural trends of the postwar period. All exhibited a marked drop in farm size and farm values, an increase in the number of farm units and in the importance of tenancy, and a decline in food output and self-sufficiency among producers.

Collectively, these changes altered the interregional diversity of the state in complicated and contradictory ways, simultaneously lessening and magnifying differences among the regions. By liquidating nearly $100 million in human property, for example, the overthrow of the peculiar institution precipitated a massive *interregional* redistribution of wealth among the state's white farm population and greatly lessened the enormous disparities in overall wealth that had separated the state's primary sections in the late antebellum period. Similarly, the agricultural reorganization that ensued – in particular, the demise of the plantation system of production that had flourished in West Tennessee – completely erased the differences in scale of operations that had previously divided the regions. On the other hand, tenancy rates in each section – which had been roughly comparable prior to the war – diverged sharply after emancipation to correlate, albeit imperfectly, with the prewar importance of slavery. The extent of self-sufficiency and commercial orientation diverged as well, as did the overall concentrations of wealth, although variations in the latter were still small and reflected the stubborn tenacity of inequality in all three regions.

Although I have chosen throughout this work to focus on the issue of interregional diversity, I believe that the examination of the early postbellum years will also prove stimulating to those concerned more with the nature and extent of southern discontinuity. If nothing else, it demonstrates unequivocally the crucial importance of careful cross-sectional and longitudinal analysis in evaluating the contours of postwar change. When applied to the Tennessee data, such systematic scrutiny yields insights that challenge at least three widely held beliefs concerning class and racial dynamics after the Civil War and the economics of postbellum agriculture.

First of all, the experience of common whites across the state – the smallholders and the landless – fails to match prominent scholarly descriptions. Although patterns of absolute economic mobility deteriorated after the war (i.e., the rate of land acquisition declined and the rate of forfeiture increased), it would be a gross exaggeration to speak of the proletarianization of white common folk during the period. The tenancy rate among whites did rise in all regions, most sharply in West Tennessee, but not primarily because marginal owners lost title to their farms. Of greater importance were a demographic trend – a booming increase in the white farm population that taxed existing resources and thwarted the aspirations of new households – and a structural trend – the shift of numerous landless whites from agricultural labor to farm tenancy.

Second, the postemancipation experience of Tennessee freedmen was markedly inconsistent with the standard scenario repeated in most accounts of the Reconstruction era. The reorganization of agriculture across the state did not lead rapidly to the predominance of sharecropping among the freedmen, nor did it preclude their frequent acquisition of land. Fifteen years after emancipation, the fate of the former slaves was actually both better and worse than conventional accounts would suggest. The vast majority of the freedmen were apparently clustered at the very bottom rung of the agricultural ladder, as wage laborers rather than croppers or tenants. At the same time, upward mobility to the highest rung of the ladder –

farm ownership – was more common than the simple comparison of ownership rates at discrete points in time would indicate.

Finally, crop production patterns in the state call into question popular views concerning the causes and extent of declining self-sufficiency among southern farmers after the Civil War. Although postbellum tenants in Tennessee were far less likely to have been self-sufficient than owner-operators, farm-specific data show that a majority of *antebellum* tenants had also been deficient in basic food production, even before the disruption of war and emancipation, the collapse of the southern credit structure, and the rise of the insidious crop lien. Although scholars have focused upon the production decisions of individual farmers to explain postbellum trends, in Tennessee, at least, declining self-sufficiency stemmed less from changes in individual market behavior than from larger impersonal forces: dwindling supplies of land and livestock relative to the farm population.

With regard to all three issues, of course, it is possible that the patterns observed in Tennessee were uncharacteristic elsewhere in the South. At present, it is simply impossible to know with any certainty. What is certain, however, is that it would be unwise for scholars to dismiss these findings until they have applied the same painstaking techniques of analysis to other southern regions. We simply do not know, for example, whether wage labor was insignificant elsewhere in the South because few scholars have systematically searched for wage laborers. Nor do we know how common it was for postbellum whites to lose their land, because few scholars have traced individual southern whites over time.

At bottom, the conclusions offered here with regard to southern discontinuity resemble those concerning southern interregional diversity. In both cases they are suggestive rather than definitive, complex, and in some ways conflicting. Because they lack the rhetorical power of bold generalizations, they are not the type of conclusions scholars generally strive to formulate, and yet I believe that they have served their purpose. They have raised questions, suggested the pressing need for future research, and demonstrated the potential rewards of an explicitly comparative approach to the study of southern heterogeneity. As scholars begin to apply such an approach, our understanding of the internal diversity of the nineteenth-century South is sure to change. Stark contrasts will undoubtedly fade, to be replaced by comparisons of greater complexity which, for all their subtlety, will be no less intriguing.

Appendix A:
Statistical Method
and Sampling Technique

Do not guess, try to count, and if you can not count admit that you are guessing.[1]

The conclusions presented in this work rest upon an analysis of extensive statistical data drawn from the federal manuscript censuses for 1850, 1860, 1870, and 1880. Although these censuses are indispensable tools for the economic and social historian, any scholar who has used them knows that they are potentially deceptive as well as insightful. To maximize their value, historians must use the quantitative evidence that they contain judiciously and with a constant awareness of their limitations – in the same manner, in other words, as they would approach any other form of historical artifact. This means applying rigorous standards both in the selection of evidence (sampling) and in its interpretation (statistical inference).

The data sets developed for this study fall into one of two basic categories. The statistics on wealth distribution presented in Chapters 2 and 3 are based on estimates of real and personal wealth recorded in the manuscript population censuses and include the *entire* population of white farm households for all eight sample counties. A farm household, as I have defined it, included any household in which one or more members reported a farm-related occupation to the census enumerator, whether planter, overseer, farmer, tenant, farm laborer, or farm hand. (The ratio of farm households to total households ranged from 0.77 in East Tennessee to 0.81 in West Tennessee.) Whenever a household contained multiple wealthholders,

1. G. Kitson Clark, *The Making of Victorian England* (London: Methuen, 1962), p. 14, quoted in William O. Aydelotte, "Quantification in History," *American Historical Review* 71 (1966):807.

I summed the figures for each member and recorded only the household total.

Unlike the evaluation of wealth distributions, the analysis of farm operations and of geographic and economic mobility relies on statistical *samples* of the relevant farm populations.[2] There are four such data sets in all, one for each census year between 1850 and 1880. Each represents a weighted composite of separate random, stratified samples drawn from all eight sample counties. With the exception of that for 1870, I drew each county sample initially from the manuscript agricultural census and stratified it according to tenure category as inferred from or reported on the schedule. I then linked each selected farmer to the population census to determine age, sex, place of birth, literacy, and size of household. For 1850 and 1860 the tenure categories employed for stratification were (1) farm-owners (those who appeared in the agricultural census and reported ownership of real estate on the population census), (2) tenants (those who appeared on the agricultural schedule but reported no real estate), and (3) farmers or farm laborers without farms (household heads who reported agricultural occupations on the population schedule but were absent from the agricultural schedule). For 1880 the tenure categories employed are those defined however imperfectly by the tenth census: (1) owner-operators, (2) operators who "rent[ed] for fixed money rental," and (3) operators who "rent[ed] for shares of products." The 1870 county samples are unique in that they were drawn initially from the population census (the 1870 agricultural census was defective) and stratified by the dichotomous variables of race and landownership. Each actually contains four discrete subsamples of white landowning households, white nonlandowning households, black landowning households, and black nonlandowning households. All four decennial samples are sufficiently large so that, when estimating the proportion of the population exhibiting a given characteristic (e.g., planting cotton or persisting geographically), the overall regional estimates should be subject to a standard error (reduced by a finite population correction factor) of 1.2 percentage points or less, producing 95 percent confidence intervals of approximately 2.2 to 2.4 percentage points.[3] The sample sizes for each year are presented in Table A.1.

2. In developing the sampling strategy, I have relied most heavily on R.S. Schofield, "Sampling in Historical Research," in E. A. Wrigley, ed., *Nineteenth-Century Society: Essays in the Use of Quantitative Methods for the Study of Social Data* (Cambridge: Cambridge University Press, 1972), pp. 146–90.

3. It is not true, as historians often implicitly suggest, that for a sample of fixed size the margin of error in estimating the mean value of a specific population item (e.g.,

Table A.1 Sample size by region, race, and tenure, 1850-1880

	East	Middle	West	Total
		1850		
Owners	703	799	563	2,065
Tenants	344	350	305	999
Farmers without farms	396	249	192	837
Total:	1,443	1,398	1,060	3,901
		1860		
Owners	691	903	473	2,067
Tenants	309	468	213	990
Farmers without farms	487	354	117	958
Total:	1,487	1,725	803	4,015
		1870		
White landowners	741	892	467	2,100
Black landowners	21	171	59	251
White landless	966	756	431	2,153
Black landless	236	697	605	1,538
Total:	1,964	2,516	1,562	6,042
		1880		
White owners	711	739	515	1,965
Black owners	56	45	75	176
White tenants[a]	104	739	515	1,965
Black tenants[a]	5	101	272	378
White sharecroppers[b]	396	453	73	922
Black sharecroppers[b]	76	208	390	674
Total:	1,348	1,961	1,452	4,761

[a]Fixed-rent tenants as enumerated on agricultural census.
[b]Enumerated as renting for "shares of crops" on agricultural schedule, but probably included both share tenants and sharecroppers.

In stratifying the 1850 and 1860 samples I was aided immeasurably by worksheets prepared by Harriet C. Owsley and Blanche Henry Clark Weaver while the latter was a doctoral student at Vanderbilt University during the 1930s. Under the direction of Owsley's husband, Professor Frank L. Owsley, the two transcribed data from the manuscript agricultural censuses of both 1850 and 1860 for every farm in eighteen (!) Tennessee counties (including those selected for this study). Significantly, because they also cross-referenced both the population and slave censuses, they were able not only to indicate on the worksheets whether farm operators owned real estate, slaves, or both but also to develop a list of farmers and farm laborers who headed households included on the population census but who were not registered on the census of agriculture. This allowed me to stratify both of the antebellum tenure samples by tenure category, a task that would have been otherwise unfeasible given that the antebellum agricultural schedules did not indicate tenure status. Because the only variable they recorded from the population census pertained to real estate ownership, I still had to search in the population schedules for each farmer sampled from their worksheets in order to determine other important personal and household information.[4]

Because I am interested in longitudinal as well as cross-sectional patterns, I also traced each household head sampled from 1850, 1860, and 1870 to the manuscript population census ten years later. When searching for individuals in subsequent censuses, I confined my examination to the county of origin and made no further attempt to locate farmers who emigrated. Enumerators' chronic inconsistency in spelling and inaccuracy in the estimation of ages often made verification of matches problematic. (This

the average number of bushels of corn harvested by farmers who grew the crop) will be less than or equal to the margin of error when estimating the proportion of the population with a given attribute (e.g., the percentage of farmers who grew corn). Because the standard deviation of a dichotomous variable is finite and that of a continuous variable is not, the margin of error when estimating a mean value will commonly be larger, frequently much larger, than when estimating the proportion with a given attribute. Far too commonly, historians quote the margin of error for the latter, even though they concentrate on issues requiring the estimation of mean values. As a result, their statistics are far less accurate than claimed. (For a typical example of this kind of mistake, see Steven Hahn, *The Roots of Southern Populism: Yeomen Farmers and the Transformation of the Georgia Upcountry, 1850–1890* (New York: Oxford University Press, 1983), pp. 291–2.

4. The worksheets are included as part of the Owsley Papers in the Special Collections Division of the Jean and Alexander Heard Library at Vanderbilt University. Mrs. Weaver reported the findings of her investigation in Blanche Henry Clark, *The Tennessee Yeomen, 1840–1860* (Nashville: Vanderbilt University Press, 1942).

is a primary reason why, in my opinion, state-wide computer-generated indexes are of limited value to anyone searching for more than a handful of individuals.) To confirm instances of persistence with greater confidence, I checked not only the name and age of the household head but also the names and ages of spouses and all dependent children residing in the parents' household in both census years. Although I did not rigidly follow any mechanical rule for verification, in general I rejected potential matches whenever the difference between the reported and "predicted" age of a possible persister exceeded five years. When the reported and predicted ages differed by five years or less, I accepted the match provided that the names and ages of other household members were approximately as expected. To lessen the possibility of oversight, I rechecked the manuscript rolls for all sample household heads not found on first examination.

To determine the absolute and relative economic mobility of persisting household heads during either the 1850s or 1860s, I merely compared the wealth values recorded in the population census at the beginning and end of the relevant decade. Because the 1880 population census contained no information on property ownership, I determined the frequency of land acquisition and dispossession by matching persisters to the agricultural schedules, which for the first time explicitly identified farmowners and tenants. In addition, I searched county tax lists for the small proportion of persisters who did not appear in the agricultural census. A tiny proportion of persisters (less than 5 percent in each county) appeared on neither the agricultural census nor local tax rolls. Although I have assumed that these individuals were landless, the aggregate figures presented in Chapters 3 and 4 would change but slightly under the alternative (and highly unlikely) assumption that such individuals were landowners.[5]

To determine the number of households headed by farmers and farm laborers without farms in 1880, I first matched the sample farmers (drawn from the agricultural schedule) to the population schedule to determine the race and reported occupation of each, thus allowing me to make overall estimates of the breakdown by race and reported occupation within each tenure category. It was then necessary to sift systematically through the

---

5. The 1880 tax records for West Tennessee's Fayette County have not survived, thus I was unable to complete this procedure for sampled farmers from that county. The result is that the frequency of land acquisition in that county has been understated, although if patterns in Fayette resembled those in Haywood, the distortion in the regional estimates reported in the text is minimal.

population schedules, recording the total number of farm households for each race. I added to the total number of farm households on the population schedule the estimated number of farm operators reporting nonagricultural occupations to arrive at an estimate of the total size of the true agricultural population. With this figure and the estimated racial and occupational compositions of the farm operators, I have been able to estimate indirectly the total number of farm households not listed on the census of agriculture, as well as their racial and occupational composition. Unfortunately, this procedure does not afford individual-level data for those farm household heads recorded in the population census only, thus preventing a more detailed analysis of the group's characteristics (e.g., age, marital status, family size).

When all is said and done and the tedious, time-consuming processes of data sampling, recording, and evaluation have been successfully completed, historians who employ quantitative evidence must take one last precaution before formulating their arguments. They must also steel themselves against the seductive powers of the numbers that their computers generate; statistics with three significant digits have mesmerizing properties that clothe oftentimes crude data with a spurious air of scientific exactitude. It is with this characteristic in mind that historian Richard Beringer relates a delightful excerpt from a report filed by a British government official, presumably early in the twentieth century:

> The Government are very keen on amassing statistics. They collect them, add them, raise them to the Nth power, take the cube root and prepare wonderful diagrams. But you must never forget that every one of these figures comes in the first instance from the village watchman, who just puts down what he damn pleases.[6]

Although the *typical* nineteenth-century census enumerator, in my opinion, did not simply put down whatever he pleased, he did labor under numerous handicaps unrelated to his personal integrity or intellectual capacity, and I have found it helpfully sobering to remind myself from time to time of the imperfect methods by which he collected his data. Even when the enumerator recorded the information reported to him with the strictest attention to detail, he was still only writing down the rough estimates of individuals not prone, as a rule, to detailed record keeping. My object is not to denigrate the value of such data – I did, after all, spend the flower

6. Quoted in *Historical Analysis: Contemporary Approaches to Clio's Craft* (New York: John Wiley & Sons, 1978), p. 198.

of my youth peering at a microfilm reader – but rather to present a realistic assessment of their precision. For all their shortcomings, the nineteenth-century censuses remain a priceless, unparalleled source of insight into the lives of anonymous Americans.

Appendix B:
Estimates of the Food Supply and the Extent of Self-sufficiency on Tennessee Farms

In determining the extent of self-sufficiency among Tennessee farm households, it has been necessary first to estimate the net production of foodstuffs available for household consumption. Net foodstuffs consist of all meat production plus all grain not fed to livestock or reserved for seeding the following year's crop. The following ratios are used in estimating the proportion of grain crops reserved for seed:

corn	0.050
oats	0.083
Irish potatoes	0.100
sweet potatoes	0.100
rye	0.125
wheat	0.125
barley	0.125
buckwheat	0.083
peas and beans	0.083[1]

The residual in each crop is then converted to corn-equivalent bushels based on the following corn conversion ratios:

corn	1.000
oats	0.433
Irish potatoes	0.220

1. Seed ratios for all crops except Irish and sweet potatoes are taken from Robert E. Gallman, "Self-Sufficiency in the Cotton Economy of the Antebellum South," *Agricultural History* 44 (1970):10. The ratios for potatoes are those suggested by both Battalio and Kagel, "The Structure of Antebellum Southern Agriculture: A Case Study," *Agricultural History* 44 (1970):25–37; and William K. Hutchinson and Samuel H. Williamson, "The Self-Sufficiency of the Antebellum South: Estimates of the Food Supply," *Journal of Economic History* 31 (1971):594–5.

sweet potatoes	0.362
rye	1.050
wheat	1.104
barley	0.866
buckwheat	0.620
peas and beans	0.946[2]

The estimated annual consumption of grain by livestock other than swine, in corn-equivalent bushels, is as follows:

horses	35.00
mules	30.00
oxen	35.00
milch cows	5.00
other cattle	2.25
sheep	0.25[3]

I assume that, in addition to grain consumed when fattening, swine consumed on average two bushels of corn from birth to selection for slaughter (typically at eighteen to twenty months of age).[4] The residual grain output, after deductions for seed and livestock feed, represents the total grain (in corn-equivalent bushels) available either for human consumption or for fattening swine.

Estimates of meat production per farm are based on the following estimated slaughter weights and slaughter ratios:

	slaughter weight	slaughter ratio
swine	—	0.445
other cattle	240 lbs.	0.286
oxen	426 lbs.	0.166
milch cows	240 lbs.	0.143
sheep	19 lbs.	0.400[5]

2. See Roger Ransom and Richard Sutch, *One Kind of Freedom: The Economic Consequences of Emancipation* (Cambridge: Cambridge University Press, 1977), p. 247.
3. With the exception of the consumption allowance for other cattle, the foregoing estimates are from Ransom and Sutch, *One Kind of Freedom*, p. 250. For estimated consumption of grain by other cattle, see Battalio and Kagel, "The Structure of Antebellum Southern Agriculture."
4. Ezra C. Seaman, *Essays on the Progress of Nations* (New York: Charles Scribner, 1852), p. 275.
5. The slaughter ratios assume an age at slaughter of twenty months for swine, two years for sheep, eight years for milch cows and oxen, and four years for other cattle. Slaughter weights and ratios are from Battalio and Kagel, "The Structure of Antebellum Southern Agriculture," p. 36.

Because estimates of the slaughter weights of swine vary so widely (and the actual averages may have varied greatly across the state), I follow Gallman and estimate potential pork production, not from estimated slaughter weights and slaughter ratios, but from surplus grain production, assuming that each bushel of surplus grain fed to swine yielded about 10 pounds of live pork or 7.6 net pounds after slaughter.[6] Unlike Gallman, however, I do not assume that pen-fed grain was the only source of sustenance for swine. Swine in Tennessee were generally range-fed until the last one to two months before slaughter, and I assume conservatively an average weight of 90 pounds per swine just prior to fattening. Thus, even on grain-deficient farms there would have been a potential pork supply equal to approximately 68 pounds – 0.76 times 90 pounds – per swine slaughtered.

Following Battalio and Kagel, the estimated minimum annual consumption of grain and meat per adult slave is placed at 1.25 pecks of grain per week and 3.5 pounds of meat; the requirement for children under fifteen is estimated to average half these amounts. Among the free population, both caloric requirements and consumption are assumed to have been lower. Following Gallman, I generously estimate per-capita consumption of grain among the free population at three-quarters of the slave level. Conversely, I assume that the free population consumed approximately 12 percent more meat per capita than the typical slave.[7] These assumptions lead to minimum annual per capita consumption requirements of 13.5 bushels of grain and 136.5 pounds of meat per slave, and 10.16 bushels of grain and 152.6 pounds of meat per free person. Expressed in corn-equivalent bushels, the total annual requirement of foodstuffs would be approximately 31.5 bushels per slave and 30.25 bushels per free individual. For the computation of self-sufficiency rates in 1880, I assume that the grain requirements of freedmen equaled the antebellum requirements of the free population, but in order to bias the results in favor of self-sufficiency among black operators, I assume that the freedmen continued to subsist at slave levels of meat consumption, rather than at the higher level estimated for the antebellum free population.

In Chapters 1 and 5 additional adjustments for tenants have also been made to reflect rent paid to the landlord. I have assumed that all antebellum

6. Gallman, "The Self-Sufficiency of the Cotton South"; Jeremy Atack and Fred Bateman, *To Their Own Soil: Agriculture in the Antebellum North* (Ames: Iowa State University Press, 1987), p. 214.

7. This is derived from Atack and Bateman's estimate of per-capita meat consumption in the rural North. See *To Their Own Soil*, p. 210.

tenants paid one-third of their grain as rent. Adjustments for postbellum tenants vary according to tenure status recorded on the 1880 agricultural schedule. The grain output of fixed-rent tenants is reduced by one-third, that of sharecroppers by one-half. Finally, because sharecroppers were typically not responsible for feeding working stock, their grain output has been adjusted accordingly.[8] The grain paid by tenants is considered a loss to tenant households, but not to the individual county, on the assumption that landlords were typically residents of the same county as their tenants.

8. Ransom and Sutch, *One Kind of Freedom*, pp. 251–2.

Appendix C:
Wholesale Price Data for Agricultural Commodities, 1859–1879

Before 1880, census enumerators did not ask farm operators to estimate the value of agricultural products produced on their farms. The estimates of production value presented in Chapter 1 are based on wholesale commodity prices quoted in Knoxville, Nashville, and Memphis (see Table C.1). Newspapers from these cities did not always quote prices for every commodity recorded in the census. Occasionally it has been necessary to employ a price quotation from one city as the statewide price. In addition, no attempt has been made to estimate the value of orchard, forest, or dairy products.

Although the 1880 census does contain estimates of the value of production, the estimates for 1879 presented in Chapter 5 are also based on wholesale price data from regional markets in order to preserve strict comparability with antebellum figures (see Table C.2).

Table C.1 Wholesale prices of selected agricultural products in
Tennessee, 1859 (dollars)

	East	Middle	West
Corn/bu.	0.75	0.675	0.775
Wheat/bu.	0.90	0.85	0.925
Oats/bu.	0.70	0.70	0.70
Rye/bu.	0.75	0.75	0.75
Sweet pot./bu.	1.125	1.125	1.125
Irish pot./bu.	0.70	0.75	0.725
Cotton/lb.	0.103	0.103	0.103
Pork/lb.	0.045	0.045	0.03
Beef/lb.	0.045	0.045	0.03

Sources: Brownlow's Knoxville Whig, 20 August 1859; [Nashville] *Union and
American*, 19 August 1859; and *Weekly Memphis Bulletin*, 9 September 1859.

Table C.2 Wholesale prices of selected agricultural products in
Tennessee, 1879 (dollars)

	East	Middle	West
Corn/bu.	0.50	0.50	0.45
Wheat/bu.	1.15	1.15	1.10
Oats/bu.	0.46	0.45	0.31
Rye/bu.	0.70	0.70	0.70
Sweet pot./bu.	0.81	0.69	0.45
Irish pot./bu.	0.725	0.69	0.45
Cotton/lb.	0.115	0.115	0.115
Pork/lb.	0.04	0.036	0.04
Beef/lb.	0.03	0.02	0.02

Sources: Knoxville Daily Chronicle, 2 December 1879; [Nashville] *Daily
American*, 2 December 1879; and *The Daily Memphis Avalanche*, 2 December
1879.

Index

FRANKLIN PIERCE COLLEGE LIBRARY

00103915